What people are saying about

All Is One

I love the messages in R   .eaning that stood out for me pers   .e to wake up. So many of us, including myse   ..s, are living life fast asleep. Daydreaming. Contemplating the time that we will get sober, find happiness, find inner peace, and find contentment. Yet, we wake up again, get on the hamster wheel of life looking forward to the weekend. So sad. So true. So sad. Did I also say it's so true? Combining scientific facts, personal experiences, and a willingness to try someone else's way… Yes, a full surrender, are also powerful messages from this book. And if you are lucky enough to be 'awake' by the time you get to the weekend, you might do something like skydiving… Or forgive someone who has hurt you… Or, as Ren writes in his book, we might take some time to look into our child or our pet's eyes and see the reality of life: they are fully living where we so desire to be, they are fully present in this moment. Grab the book. Now. But read it, very, very slowly. Mindfully. It's time to change your life. **David Essel, M.S., O.M.**, Counsellor, and author of the number one best-selling book: *Focus! Slay your goals.* https://davidessel.com

All I can say about *All Is One* is WOW, what a great book. As a person who has been in the world of recovery for 30+ years, it goes straight to the heart of the 'missing piece' for most alcoholics and addicts. It is a spiritual malady and as humans we scream out to fill that spiritual part of us which in these modern times can be difficult to achieve. Some people may think spirituality and science are at odds with one another – who says SO?!?!?! They are both important and both are needed

to make the most of this human experience. Thank you, Ren, for pulling this together in such an eloquent and meaningful way.
The Addiction Doctor, Dr Robb Kelly
https://robbkelly.com

Ren allows us to see the complexity of consciousness as a single thread connecting spirituality, philosophy, religion and science, via a personal and honestly written perspective. There is something in this book that we can all learn from, which I imagine may vary for all of us. *All Is One* shines a light on everyday consciousness where you're not just accepting it but questioning it. If you can find yourself overwhelmed by the state of the world, lose yourself in these pages and you might just discover a connection within all of it laid bare before your eyes. Each chapter invites you to explore a new realm of consciousness for yourself. Whatever your beliefs, it's easy to appreciate the patterns of connectedness observed and shared here by Ren. Insightful and thought provoking – I highly recommend the read!
Poet, Jyoti Devi

I have had the pleasure of interviewing Ren on my *Funky Brain Podcast*, and I have also been on his podcast, *Life in Recovery*. We have become great friends even though we live 5,000 miles apart. That is because we share a common purpose, which is to help other people struggling from alcoholism and addiction, but ultimately to achieve 'Emotional Sobriety'. Ren has humbly dedicated his life to this and he is a fine example of somebody who embodies altruistic principles. That being said, his 5th book, *All Is One*, is the perfect tool for anybody looking to enhance their spirituality, or for someone who isn't quite sure what's out there, but is a seeker and looking for answers. I must admit, when I started reading it I thought it was a little deep for me, but the way he ties addiction in with science and spirituality is brilliantly enlightening and encouraging. My favourite concept that he draws upon is how, throughout history, humans have created hundreds of different religions and sects which have influenced billions of people (for better or for worse)... but ultimately, in the end, the truth is that there is only one underlying reality, which leads to the title of this masterpiece: *All Is One*. This book is full of carefully researched facts and theories about addiction, psychedelics, recovery principles, and of course, meditation, God and spirituality that will satisfy even the greatest critics, but is also grounded in Ren's worldly experience as a traveller, athlete, DJ, and most recently, a family man. So, anybody can relate to his story and apply the principles he is sharing into their lives with ease. This is a great read. Highly recommended. Five Stars.

Life Coach and Host of the *Funky Brain Podcast*, **Dennis Berry**
https://dennisberry.com

Previous Books

Addiction Prevention: Twelve Steps to Spiritual Awakening
ISBN: 9781543293395

Anonymous God: Coincidence, Serendipity, Synchronicity,
Spiritual Signposts and Psychedelics
ISBN: 9781985091214

Together: An Ayahuasca Experience
ISBN: 9781797717180

The Spiritual Malady: How to Attain Peace of Mind and
Lasting Happiness
ISBN: 9798612262251

All Is One

The Science & Spirituality
of Consciousness

All Is One

The Science & Spirituality of Consciousness

Ren Koi

BOOKS

Winchester, UK
Washington, USA

JOHN HUNT PUBLISHING

First published by O-Books, 2021
O-Books is an imprint of John Hunt Publishing Ltd., 3 East St., Alresford,
Hampshire SO24 9EE, UK
office@jhpbooks.com
www.johnhuntpublishing.com
www.o-books.com

For distributor details and how to order please visit the 'Ordering' section on our website.

ISBN: 978 1 78904 868 1
978 1 78904 869 8 (ebook)
Library of Congress Control Number: 2021930560

A CIP catalogue record for this book is available from the British Library.

Design: Stuart Davies

UK: Printed and bound by CPI Group (UK) Ltd, Croydon, CR0 4YY
Printed in North America by CPI GPS partners

We operate a distinctive and ethical publishing philosophy in
all areas of our business, from our global network of authors to
production and worldwide distribution.

Contents

I dedicate this book to my daughter Alice,
my wife Adele, my stepchildren Lillian and Isobel, and
my dog Teddy.

In loving memory of Hannah Kniveton.
Gone but never forgotten... her consciousness lives on.

Foreword

The journey to enlightenment is really the only quest that matters in life, but most people are asleep. As children, we are raised to survive in the world but these teachings often carry within them the seeds of dysfunction. Self-centredness, self-sufficiency and self-righteousness are at the root of our problems. We fall into the trance of seeking what we think we want, and lose sight of the greater truth: that we are all one.

Through the deconstruction and analysis of a variety of scientific studies, philosophers and thought leaders, Ren offers conclusions that will help you connect the dots on some very complex ideas. These ideas make the case that, as we understand consciousness, we can understand our place in the world and how we affect one another.

The world is full of suffering and delusion, however, there is hope and meaning through understanding. There are many ways to discover deeper meaning as suggested throughout this book, from ancient traditions to the use of alternative medicines. Ren's personal experiences and wisdom offer insights to what is possible for all of us who are seeking to unravel the conditioning of our past experiences, so that we might come to a new understanding and a new experience of possibility and awareness.

At the core of *All Is One* is courage. The kind of courage required to challenge our own self-made prison of thoughts and beliefs. Courage to bring our bias, assumptions and limited thinking into the light where all things are possible. We all have periods in our lives where we self-sabotage and get stuck, but having the courage to live with an open mind is the path to ascending to higher levels of consciousness available to all of us.

Ren was such a great guest on my podcast that we have

continued to collaborate together to help spread the message of recovery. His thoughtful insights and questions help those who listen to see their own potential and inspire to give hope for their own sobriety.

Reading this thoughtful book, you will go on a journey investigating the bigger questions, the deeper questions, that we all need to consider as we walk toward finding meaning in our lives. Through the exploration of what scientists and philosophers have to say about consciousness, before drawing your own conclusions, this book will expand your mind... and your consciousness.

Arlina Allen

Host of the *ODAAT Chat Podcast*

https://odaatchat.com

Acknowledgements

I'm very grateful to my parents Jenn and Tom Loxley, Adele, Alice, Lillian and Isobel, Teddy, Ann and Len Kniveton, Arlina Allen, Dennis Berry, Dr Robb Kelly, David Essel, Jyoti Devi, Jonathan and Jennifer Bagshaw, Shaun and Michelle Marriott, Adam McGuire, Pete Martin, Carl Jenkins, Melissa Fewtrell-Graham, Alex M., all my fellows in Twelve Step fellowships and all my family and friends.

Peace and love,

Ren

Introduction

René Descartes, a famous philosopher from the 17th century, once stated, "I think, therefore I am." These were the pioneering words that pushed humanity forward into the abstract world that is the concept of consciousness, which, over three centuries later, still remains a mystery. Consciousness involves the subjective experience of phenomena – a kind of epiphenomena, or je ne sais quoi experience that we cannot measure. Neuroscience, for example, cannot teach a blind person what it is like to see or to experience the colour blue. The purpose of this book is to discuss what the 'am' is that Descartes was referring to, which gives rise to felt experience – and how it can be explained, as we are certainly not our thoughts. Any conscious system that can observe its own thoughts, as humans can, is by definition something apart from its thoughts – and this is the subject of much scientific and spiritual investigation and debate. I would argue that a more appropriate way to frame consciousness is: "I am, therefore I think."

'Dualism' is the notion that physical and mental phenomena are somehow irreconcilable; two completely different things. According to Descartes, dualism concerns substances, as the body is made of physical 'stuff', while the mind is made of mental 'stuff'. Thanks to advances in both physics and biology, nobody takes substance dualism seriously anymore. The alternative is something called 'property dualism', which acknowledges that everything – body and mind – is made of the same basic stuff (quantum phenomena such as quarks), but that this stuff somehow changes when things get organised into brains, and special properties appear that are nowhere else to be found in the material world. The 'illusionists', by contrast, think that everything is made of the same basic kind of 'stuff', and that there are no special barriers separating physical from mental phenomena. However, since the illusionists agree with the dualists that phenomenal consciousness seems to be rather spooky, they simply deny the existence of whatever appears not to be physical and call it an illusion.

Illusionism was labelled, "The silliest claim ever made", by British philosopher, Galen Strawson, in the *New York Review of Books* in 2019, but is defended by other prominent philosophers, particularly by American philosopher and cognitive scientist Daniel Dennett. Dennett is arguably the one who began this trend back in the early 1990s with the publication of his influential book *Consciousness Explained*, which, though certainly interesting, did not explain consciousness. Dennett suggests that phenomenal consciousness is a 'user illusion' akin to the icons we use on our computer screens. He explains it like this: "When I interact with the computer, I have limited access to the events occurring within it. Thanks to the schemes of presentation devised by the programmers, I am treated to an elaborate audio-visual metaphor, an interactive drama acted out on the stage of keyboard, mouse, and screen. I, the User, am subjected to a series of benign illusions: I seem to be able to move the cursor (a powerful and visible servant) to the very place in the computer where I keep my file, and once that I see that the cursor has arrived 'there', by pressing a key I get it to retrieve the file, spreading it out on a long scroll that unrolls in front of a window (the screen) at my command. I can make all sorts of things happen inside the computer by typing in various commands, pressing various buttons, and I don't have to know the details; I maintain control by relying on my understanding of the detailed audio-visual metaphors provided by the User illusion."

This is a very powerful (metaphorical) description of the relationship between phenomenal consciousness and the underlying neural 'machinery' that makes it possible, but the term 'illusion' brings to mind deception, which is most definitely not what is going on. Computer icons and cursors are not illusions, they are effective representations of underlying machine-language processes. It would be far too tedious for most users to think in those terms, and way too slow to interact with computers via machine-language. This is why programmers gave us icons and cursors. If they were illusions, nothing would happen – they would be completely ineffective. The feelings and thoughts that we have are high-level representations of the underlying neural mechanisms that make it possible for us to perceive, react to, and navigate the world (entirely different from icons). Instead of clever programmers, we can thank billions of years of evolution by natural selection for these neural mechanisms that allow us to have conscious experiences. However, it is certainly true (as the illusionists maintain) that we do not have access to our own neural mechanisms, but we don't need to – just as a computer user doesn't need to know machine-language. This does not imply that we are somehow mistaken about our thoughts and feelings any more than I, when using my computer, might be mistaken about which folder contains the file on which I have been writing this book.

Philosopher and proponent of 'panpsychist' (the strongest opposing belief to illusionism, which states that everything material, however small, has an element of individual consciousness), Dr Phillip Goff, wants us to consider how "the father of modern science", Galileo, created The Consciousness Problem when he separated quantitative information from qualitative. By leaving qualitative information out of scientific inquiry, Galileo introduced the mind-body dualism we are still wrestling with today. Panpsychism is Goff's proposed scientific solution. Galileo thought that there was a mathematical language embedded in the cosmos that could only be seen once qualitative phenomena were removed from the quantitative, which was an idea first introduced by Ancient Greek philosopher, Plato. English mathematical physicist and philosopher of science, Sir

Roger Penrose, wrote in *The Emperor's New Mind*, "I imagine that whenever the mind perceives a mathematical idea, it makes contact with Plato's world of mathematical concepts... When one 'sees' a mathematical truth, one's consciousness breaks through into this world of ideas, and makes direct contact with it... When mathematicians communicate, this is made possible by each one having a direct route to truth, the consciousness of each being in a position to perceive mathematical truths directly, through this process of 'seeing'... Since each can make contact with Plato's world directly, they can more readily communicate with each other than one might have expected. The mental images that each one has, when making this Platonic contact, might be rather different in each case, but communication is possible because each is directly in contact with the same externally existing 'Platonic world!'" In Galileo's observations, he therefore removed sensory data derived from the five senses, and was left with a set of quantitative data – size, shape, location, motion – that became the basis for a new paradigm called science, which went beyond the limits of philosophical reasoning to the development of the scientific method.

If we are to solve The Consciousness Problem laid out by Galileo, I feel that philosophical reasoning and the scientific method will have to join forces. I am very excited by the possibility that the desires of philosophy and spirituality, and the rationality of science, might finally fall into harmony. I believe panpsychism is the mediator between dualism and materialism, and as a result, I feel we are on the verge of a new paradigm – whereby we will experience the scientific validity of panpsychism and the mind-body dualism problem will be resolved once and for all. Once we universally recognise that human consciousness is not the centre of the universe, rather, consciousness abounds, we will experience a joint scientific-spiritual quantum shift of the same magnitude the world experienced when it was proven (in 1725) that our Earth is not

All Is One

the centre of the solar system (originally proposed by Polish mathematician and astronomer, Copernicus, in 1514). I believe it will one day become common knowledge that consciousness is an intrinsic property of matter. When you think about it objectively, it's the only intrinsic property of matter that we know for certain; we know it directly as material conscious beings. All the other properties of matter have been discovered by mathematical physics and this method of discovering the properties of matter means that only relational properties of matter are known, not intrinsic properties. This is why philosophy and spirituality are equally as important as science in explaining consciousness.

Plato suggested that behind the veil of reality lies the 'truth' in 'Ideas and Geometric Forms', whereby the 'Form of The Good' (God or consciousness) is the superlative. In his dialogue entitled *Timaeus*, Plato described the 'dēmiourgos', which was a common noun meaning 'craftsman', but gradually came to mean 'producer', then eventually 'creator'. In the Platonic schools of philosophy, the 'demiurge' is an artisan-like figure responsible for fashioning and maintaining the physical universe. In most neo-Platonic systems, the demiurge is considered the fashioner of the universe but is not itself 'The One', or the 'Form of the Good', as outlined in his Socratic dialogue *The Republic*. It is theorised that the first Western philosopher, ancient Ionian Greek Pythagoras, who was a contemporary of Plato, travelled to India and trained with Sadhus in the art of samadhi meditation. It is proposed that he not only discovered through meditation that in a right-angled triangle the square of the hypotenuse is equal to the sum of the squares of the other two sides, but also that music had mathematical foundations. Meditation also led him to his theory of 'metempsychosis' (the transmigration of souls), which holds that every soul is immortal and, upon death, is reincarnated into a new body.

I personally believe that consciousness is the 'creator' or

'producer', it is the origin of all things, is forever active in the world of nature, and introduces intuition, creativity, relationships, reasoning, and imagination into a humanly constructed world that is becoming ever-increasingly divorced from nature. In the argument between naturalists such as philosopher Ernest Nagel and idealists such as Plato, I don't think it is so black-and-white, as I don't see any reason why the phenomenon of consciousness couldn't be 'tuned in' by the brain and other vital organs in the body (such as the heart), in the same way that a radio receives a signal. Emotions manifesting from consciousness via the five senses and thoughts manifesting from emotions is, in my opinion, a perfectly plausible explanation of what it is to be a human.

I also believe that becoming progressively more conscious and self-aware, by seeking fulfilment and change through personal growth, is the precursor to 'self-actualization', as American psychologist Abraham Maslow famously termed it. Once one has realized their physiological, safety, love/belonging and esteem needs, and begun to practise morality, creativity, spontaneity, problem solving, lack of prejudice and acceptance of facts (see image below*), as I did from 2010 to date, one is free to become everything one is uniquely capable of becoming – and able to move toward total liberation from all suffering; enlightenment.

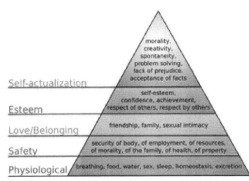

The pyramid has become synonymous with Maslow's hierarchy of needs despite Maslow never actually using a pyramid.

Following my spiritual awakening in 2009, when I hit 'rock-bottom' as a result of alcoholism, and lots of hard work in recovery, I reached a place where I was free from the insanity of obsessive-compulsive thinking and behaviour patterns. I progressively became free from dishonesty, free from isolation, free from fear, and free from the bondage of self, as my ego-mind had loosened its grip. To achieve this, I worked through the Twelve Step Program with a sponsor numerous times on various addictive patterns including alcohol, sex, and food, and I continue to practise and incorporate the Twelve Step principles of love and service in all my affairs today. I also had years of person-centred counselling and psychotherapy, participated in psychedelic therapy sessions with ayahuasca and psilocybin, completed a ten-day silent meditation retreat (where I learned to practise Vipassana meditation and mindfulness), and most recently worked with a life coach, which helped me to focus on one goal at a time. The work is ongoing.

This book is an exploration of consciousness via the topics of science and philosophy, religion and spirituality, psychedelics, drugs (including alcohol), recovery, God, meditation and enlightenment. It is also a personal account of how it feels to live life in the 'now' as a self-actualized person – in recovery from my ego-mind – and becoming progressively more liberated from my attachment to material things, destructive behaviour patterns, and thoughts that no longer afford me any sense of comfort or contentment. Peace of mind, and the resulting happiness it brings, was always my ultimate goal. Having achieved this and learned to live a contented life without having to reach for all the things I used to think would make me happy – drugs (including alcohol), casual sex, and food – true joy has consistently arisen. I mostly find myself in a general state of contentment, and I've discovered there is no greater joy than being present with my family.

I also enjoy an informed apathy toward the existence or nonexistence of any gods or goddesses, angels or demons, because I understand that enlightenment is instituted by wisdom, which is grounded in knowledge, and knowledge is founded in scientifically verified reality, which is shifting all the time. Often, what was considered scientific fact a few centuries ago can now be seen as allegory, such as the Earth is flat or the Earth is the centre of the universe. Conversely, archaeological evidence suggested that the oldest civilizations were approximately 5,000 years old, but in recent history this myth has been dispelled, as advanced civilizations older than 12,000 years have been discovered. I talk about this in more detail in Chapter 6. The point being that I don't believe anything unless it is scientifically proven, or unless I've experienced it for myself – and I've had many seemingly inexplicable and supernatural experiences that, when I boil it down, can be explained by the simple fact that everything is interconnected by consciousness.

Despite my personal choice not to include religion in my consistent worldview, I have bags of faith – and knowledge and understanding have revealed to me that: All Is One. The One of which I speak is consciousness, and consciousness goes by many different names: God, Higher Power, Tao, Brahman, the zero-point field (ZPF). Anything that has the power to bring matter (particles and waves) in and out of existence is what I refer to as a Higher Power. I named my own personal Higher Power The Organizing Principle (TOP), which is an abstraction, sounding more like a code, formula, sequencer, or enigma. This makes sense to me because consciousness organises the physical universe and it is the principal truth that serves as the foundation for all matter and sentient life. Also, the word 'consciousness' sounds like a 'thing' to me and it is clearly not a thing, in fact it is actually 'no-thing', but I get into that in more detail in Chapter 6.

When I began my recovery from alcoholism in 2010, it was suggested that I find a Higher Power of my own understanding that would solve my problem. Through scientific research, philosophy, and spiritual practices and experiences, I found a Higher Power that I now understand to be consciousness, and it *has* solved my problem. I enjoy peace of mind the majority of the time and I am getting progressively happier and content as time goes by – despite the ups-and-downs of life. I hope, therefore, you can enjoy reading about how I came to the wisdom that has afforded me contentment and peace of mind as much as I enjoyed writing this book for you.

Chapter 1

Science & Philosophy

When my daughter Alice was born in February 2020, my little bundle of joy got me thinking even more about consciousness than usual, which led to the creation of this book. When I held Alice in my arms for the first time, I was completely present. There were no thoughts of past or future, there was only moment-to-moment awareness. The feeling was one of calm, peace, serenity. That feeling stayed with me for the rest of the day, and later that evening, when I was walking my dog Teddy near the hospital (while my parents visited their granddaughter), I felt liberated from my ego-mind due to being present in the now. It wasn't long, however, until that feeling passed and the next day I was again living life from my ego-mind, flitting from past to future, which is the antithesis of living in the now.

Research shows that the more time you spend mind-wandering (not being present) the less happy you are likely to be. In fact, humans devote a whopping 47% of their waking hours to mentally time-travelling, so it's no wonder many people don't feel happy a lot of the time. Fortunately, you can develop present-moment awareness through practising mindfulness and meditation. Recent advances in two key scientific areas of study – neuroimaging and neurochemistry – have highlighted the measurable brain changes that result from practising meditation and lead to improved cognition and mood. Meditation is inextricably linked to consciousness and it is the subject of Chapter 7, so I'll get back to my original point.

Sat in my favourite armchair at home, a few days after Alice's birth, I studied my baby's face while holding her in my arms. I saw myself in her and I truly considered, for the first time, that I was once a tiny helpless baby that my parents had managed to keep alive long enough for my ego to manifest, for my character to develop, and for me to become my own person. Every time I stared into Alice's wide turquoise-blue eyes, I noticed there wasn't much going on – other than her attempting to observe what was right in front of her face. "Why would there be anything more going on?" I asked myself rhetorically.

Every newborn baby is completely present, with no sense of past or future. They are only aware of their immediate needs: attention, food, comfort and love. It's scientifically valid to say that they have no sense of self at all. One might argue that consciousness is simply *experiencing* the organism (and the world around it) rather than *controlling* the organism. A baby does not decide what to think or feel any more than he/she can decide what to see or hear, and this truth continues throughout a person's life. None of us have any control over any of the conscious or subconscious processes that happen within us or to us. Consciousness, therefore, is a witness to the unfolding of events but I think it is also true to say that consciousness is the first cause of all events, as consciousness seems to be a mysterious combination of energy and information that gives rise to all thoughts, feelings and matter. According to American theoretical physicist, John Archibald Wheeler, "The universe is fundamentally an information-processing system from which the appearance of matter emerges at a derivative level of reality."

It is interesting to note that babies even think they are one with their mother (life-giver) until around four months of age, and that feeling of 'oneness' is what can ultimately liberate us from suffering in later life. I think it's logical to conclude that every baby is born 'enlightened', which is why enlightened masters continuously advise us to be more child-like. Their joy abounds at all times, unless they are hungry or tired!

As we grow from a tiny, helpless baby into adulthood we become completely disconnected from that feeling of 'oneness' because we identify only with our name and our body. We use substances and behaviours to numb ourselves from feelings and we rarely sit in meditation, boredom, or contemplation because we are always on the go. Conversely, when you were a newborn baby, you did not know who you were and you did not identify with a name. A baby eats when they are hungry rather than snacking on sweet treats in-between meals. They cry when they are upset rather than bottling up emotions, and they sit happily observing their surroundings, perfectly meditative, with no need to distract themselves from the present moment with the next 'fix'. This is how humans are supposed to exist in the world, fully conscious in the now, present and aware, but it all starts to go horribly wrong around two years of age when the ego begins to develop – and this is why we call it the 'terrible twos'. The ego wants to take credit for every action that happens to, and inside, the body in order to get appreciation – but the ego can never be satisfied. Somehow, we must learn to mitigate the ego's incessant demands and live a 'conscious' life if we are to be truly happy.

Isolating at home from March 2020 due to the COVID-19 pandemic, I was not only blessed with lots of quality time with my newborn daughter, I also had plenty of time to research the latest science and spirituality regarding consciousness – and to contemplate how this most important and pressing subject relates to my own life and the life of every human on the planet. Ever since I left school in 1997, I've maintained an interest in science, as science is about experimentation followed by the explanation of the findings of experiments. It informs us what 'reality' is and I like to be informed about reality. When people thought the Earth was flat, science proved it was not flat through experimentation then explained the reasons why it is not flat. I like science but science doesn't always get it right, and science is subject to change, especially when it comes to reality. So, what does science have to say about consciousness?

After hours and hours of research, I could not accept the

mainstream reductionist and naturalistic view, espoused by the likes of American philosopher Daniel Dennett, that consciousness is simply emergent phenomena of the brain. And I certainly do not agree with the late theoretical physicist, cosmologist, and author, Stephen Hawking, who famously stated, "I think the brain is essentially a computer and consciousness is a computer program. It will cease to run when the computer is turned off." The first question that springs to mind is: who created the program? To believe, as Hawking did, that, "the human race is just a chemical scum on a moderate-sized plant, orbiting around a very average star... We are so insignificant that I can't believe the whole universe exists for our benefit", suggests there is no meaning, purpose or value to human life, which I cannot reconcile due to my vehemently opposing life experiences. I personally feel that the sole meaning of life is to live a life of meaning, and the life of every organism is meaningful simply because it exists. The odds of manifesting as a human being are 400 trillion to one, which is unfathomable really. Once you come to appreciate the odds of you being here now, you can be truly grateful for the miracle of your existence. It not only makes me grateful for my own existence but especially grateful for the existence of my daughter.

The current scientific model of the material world obeying laws of physics is so dominant that we often forget about our starting point as conscious observers and nonchalantly conclude that matter is the only reality. However, we are in fact substituting the reality of our experience with belief in an independently existing material world that gives rise to conscious experiences. It seems far more plausible to me that consciousness has its own properties, as quantum mechanics is beginning to suggest, and neglecting this leads us to a fundamentally incomplete description of the universe. I would also go one step further to argue that it is consciousness, as a creative agent, that gives rise to matter rather than the other way around.

I agree with the Buddhist view that the physical world, as we experience it, emerges from a realm of universal, archetypal forms, which evolves from a subtler ubiquitous dimension known as the formless realm or 'aether' in Western transcendental mysticism. In the case of the 'Big Bang' theory there exists an opposing, far more realistic but less popularised theory known as the Steady State universe. Steady State posits the continuous creation of matter throughout the universe to explain its apparent expansion. This type of universe would most likely emanate from a ubiquitous 'source field' and be infinite, with no beginning or end, which makes far more sense to me than a universe that began with every speck of its energy jammed into a tiny point. We are then led to believe this extremely dense point exploded with unimaginable force, creating all matter and propelling it outward to make the billions of galaxies of our vast universe, and then gave rise to consciousness.

Of all the scientific theories of consciousness I researched, the stochastic electrodynamics (SED) framework, a branch of physics that sheds light on the basic principles underlying quantum systems, offers the most satisfying explanation in my opinion. SED is based on the theory that all conceivable shades of phenomenal awareness are woven into the frequency spectrum of a universal background field, named zero-point field (ZPF), which is a ubiquitous ocean of zero-point energy. This "field of infinite possibilities" or "possibility field" (as Indian-American author, Deepak Chopra, terms it) implies that the fundamental mechanism underlying conscious systems rests upon access to information available in the ZPF (often referred to as the Grid, the Matrix, the Higgs Field, and the Akashic Records in Hinduism). The ZPF has zero-point energy (ZPE), which is the lowest possible energy that a quantum mechanical system may have, and it can bring particles in-and-out of existence. The empty space of a vacuum also has these creative properties.

According to Quantum Field Theory, the universe can be thought of not as isolated particles but continuous fluctuating fields: matter fields, whose quanta are fermions (leptons and quarks), and force fields, whose quanta are bosons (photons and gluons). These fluctuating zero-point fields lead to a kind of reintroduction of an 'aether' in physics, however, this aether cannot be thought of as a physical medium. In the simplest terms, vacuum states, zero-point fields, can spontaneously create energy, and electromagnetic radiation can be transformed into matter and vice versa. Also, as there is no natural candidate for what might cause what we call 'dark energy', the current best guess is that it is the zero-point energy of the vacuum. This problem, known as the Cosmological Constant Problem, is one of the greatest unsolved mysteries in physics.

This theory revolutionizes our notion of reality by giving significance to the ZPF as a creative agent that shapes matter and is the root cause of quantum phenomena. One of the key insights from SED is that quantum phenomena are emergent phenomena that can be traced back to the homogeneous and isotropic background field. This scientific explanation perfectly jives with my spiritual belief that an ever-progressing complexity of matter emanates from a universal background field of pure energy-consciousness. Swiss psychoanalyst, Carl Jung, and Austrian theoretical physicist, Wolfgang Pauli, referred to it as 'Unus mundus' (the concept of an underlying unified reality from which everything emerges and to which everything returns meaning 'one world' in Latin) and religious people most commonly refer to as 'God'.

Every material system (including humans) can be regarded as an open, random system in permanent contact with the ZPF. According to research from the Department of Consciousness Research (DIWISS) in Germany, evidence suggests that the brain is receptive to the flow of ZPF information that constitutes the recording of conscious experiences, suggesting that our sense of self, and the retrieval of memories, is accomplished by periodically reading and filtering persistent information states from the ZPF. The data also supports the conclusion that meditative practices and the use of psychedelic drugs/medicines detune the filter, thus preventing self-referential consciousness, which leads to the dissolution of the ego. Instead, the brain taps into a wider spectrum of ZPF modes and gains access to an extended phenomenal 'colour palette', resulting in expanded consciousness. Having experienced these expanded states of consciousness through meditation (a ten-day Vipassana retreat in 2018) and psychedelics (an ayahuasca retreat in 2017 and psilocybin in 2019), I wholeheartedly concur with the findings of this research.

SED answers the question: how does felt experience arise out of seemingly non-sentient matter? The Australian philosopher David Chalmers famously termed this the 'hard problem' of consciousness in his now classic paper, "Facing Up to the Problem of Consciousness", and further explored in his 1996 book, *The Conscious Mind: In Search of a Fundamental Theory*. Unlike the 'easy problems' of explaining behaviour, or understanding which processes in the brain give rise to various functions, the hard problem lies in understanding why some of these physical processes have an experience associated with them at all.

American psychologist, Gregg Henriques, believes we should think about the hard problem as two different problems, rather than one. We can call the first problem the 'ontological' problem – the common term for which is the 'neuro-binding' problem. This is the theory of how the brain actually produces the first-person experience of being. When we consider that much of the neuro-information processing which goes on in our brain is nonconscious we can ask, what is the magic ingredient that turns the light of experience on? and hypothesise, how might the brain be producing conscious experience, enabling us to feel things? The second 'hard problem' is the 'epistemological' problem. This pertains to the fundamental difference between 'first person' and 'third person' viewpoints on knowing. It works as follows: A third-person viewpoint is a view that can be taken by an external observer. An easy way to think of a third-person view visually is that it is anything that can be captured by a video camera. In contrast, the first-person view is the view behind your eyes. This is fully contained within the individual and of course cannot be filmed by a camera. This containment results in two important epistemological difficulties, which are mirror images of each other. Henriques calls this the 'epistemological gap' because it relates to how we can know what we know. The first is the problem of directly knowing another's subjective experience – the problem being it cannot be done. This is the problem of, "How do I know that you see yellow the way I see yellow?" This problem also relates to our knowledge of consciousness in other animals, which we can only know indirectly. This is the point that American philosopher, Thomas Nagel, makes in his famous 1974 paper, "What Is It Like to Be a Bat?" The second issue is that, as individuals, we are in some ways trapped in our subjective perceptual experience of the world. The only way we can know about the world is through our own subjective experiences. What it does mean, however, is that the unique experience of being-in-the-world for each of us as particular individuals is an 'extra-scientific' domain. Meaning that our idiographic experience of being resides outside of the purview of scientific knowledge. It is important to note that there are other important 'extra-scientific' domains, such as questions of ethics and morality. Science builds models about the behaviour of the universe across different dimensions and levels of analysis but science does not tell us how we should behave, nor does it give us a definitive theory of the unique, individual experience of being-in-the-world from a first-person perspective. Indeed, science struggles to do this both ontologically and epistemologically.

Theoretical physicist Albert Einstein stated, "No problem can be solved from the same level of consciousness that created it." Therefore, it might take a quantum-leap in human consciousness before we can truly understand consciousness. The fact that the hard problem has persisted for so many decades, despite the advances in neuroscience, has caused some scientists to wonder if they've been thinking about the problem backward. Rather than consciousness arising when non-conscious matter behaves a particular way, they began asking, is it possible that consciousness is an intrinsic property of matter – that it was there all along?

According to Dutch computer scientist and philosopher Bernardo Kastrup, "Consciousness cannot have evolved... The sooner we acknowledge it, the sooner we'll solve the hard problem of consciousness." In his February 2020 essay for iai news Kastrup stated, "Under the premises of materialism, phenomenal consciousness cannot – by definition – have a function. Our phenomenal consciousness is eminently qualitative, not quantitative. There is something it feels like to see the colour red, which is not captured by merely noting the frequency of red light. Experiences are felt qualities – which philosophers and neuroscientists call 'qualia' – not fully describable by abstract quantities. But, qualities have no function under materialism, for quantitatively-defined physical models are supposed to be causally-closed; that is, sufficient to explain every natural phenomenon. As such, it must make no difference to the survival fitness of an organism whether the data processing taking place in its brain is accompanied by experience or not: whatever the case, the processing will produce the same effects; the organism will behave in exactly the same way and stand exactly the same chance to survive and reproduce. Qualia are, at best, superfluous extras. Therefore, under materialist premises, phenomenal consciousness cannot have been favoured by natural selection. Indeed, it shouldn't exist at all; we should all be unconscious zombies, going about our business in exactly the same way we actually do, but without an accompanying inner life. If evolution is true – which we have every reason to believe is the case – our very sentience contradicts materialism. This conclusion is often overlooked by materialists, who regularly try to attribute functions to phenomenal consciousness... All conceivable cognitive functions can, under materialist premises, be performed without accompanying experience... The impossibility of attributing functional, causative efficacy to qualia constitutes a fundamental internal contradiction in the mainstream materialist worldview. There are two main reasons why this contradiction has been accepted thus far: first, there seems to be a surprising lack of understanding, even amongst materialists, of what materialism actually entails and implies. Second, deceptive word games – such as that discussed above – seem to perpetuate the illusion that we have plausible hypotheses for the ostensive survival function of consciousness. Phenomenal consciousness cannot have evolved. It can only have been there from the beginning as an intrinsic, irreducible fact of nature. The faster we come to terms with this fact, the faster our understanding of consciousness will progress."

In February 2020, Mind Matters News reported on Kastrup's response to biologist Jerry Coyne's claim that consciousness just happens to exist as a by-product of an evolved, useful trait that promotes survival in humans. Kastrup reportedly said, "Properties that perform no function cannot have been favoured by natural selection... I can imagine that some trivial and relatively low-cost (in terms of metabolism) biological structures and functions could be merely accidental, but the brain's wondrous putative ability to produce the qualities of experience out of unconscious matter is anything but trivial. Indeed, it is nothing short of fantastic, the most stunning thing physicalists claim, the second most important unsolved problem in science according to *Science* magazine; and now it is a by-product?! Physicalists have no idea – not even in principle – how the material brain could possibly produce experience. Therefore, they appeal to – and hide behind – the inscrutable complexity of the brain with all kinds of promissory notes. Phenomenal consciousness – they argue – is somehow an emergent epiphenomenon of that unfathomable complexity. Fine. But if such is the case, it becomes unreasonable to posit that something requiring such a level of complexity could have been just an accidental by-product of something else. One can't have it both ways."

Suggesting consciousness is an "accidental by-product" and even denying that consciousness is real seems to be the best materialists can offer because the materialism they are married to makes no sense in terms of consciousness. In my opinion, Kastrup rightly asserts that consciousness cannot have evolved. Evolution deals with things that can be measured quantitatively but consciousness cannot be quantified. Similarly, I feel that British physicalist David Papineau, who says consciousness is just "brain processes that feel like something", is wrong in this assumption. Papineau believes that conscious states are just ordinary physical states that happen to have been co-opted by reasoning systems. He states that consciousness doesn't depend on some extra shining light, but only on the emergence of subjects, complex organisms that distinguish themselves from the rest of the world and use internal neural processes to guide their behaviour. The only reason that many people feel there is a 'hard' problem in his opinion, is that they can't stop thinking in dualist terms. They have a strong intuition that the brain is one thing, and that the conscious feelings are something extra – like a force field that floats above the physical matter of the brain. Then, he believes, "we do have a problem; if only we could stop ourselves seeing things through dualist spectacles, we'd no longer feel that there is anything puzzling about consciousness."

Thinking of consciousness as a fundamental field that is everywhere and imbues all matter is a more realistic explanation in my opinion. While panpsychism has been attached to a wide range of ideas throughout history, contemporary panpsychism describes reality very differently than the earlier versions, and it claims to be unencumbered by any religious beliefs. Modern panpsychism is informed by the sciences and fully aligned with physicalism and scientific reasoning. It is due to panpsychism and this definition of consciousness given by Thomas Nagel – "An organism is conscious if there is something that it is like to be that organism" – that I firmly believe computers will become progressively super-intelligent but never conscious. Consciousness determines whether there is an experience present or not in an organic system, therefore, consciousness is not linked to intelligence but is linked to organic biochemistry in such a way that it will never be possible to create consciousness in non-organic systems.

A combination of SED and panpsychism, science and spirituality, explains my worldview, whereby consciousness is essentially a 'Higher Power' that brings all things into creation. As consciousness goes to higher and higher levels in humans through mechanisms such as meditation, one's relationship to the body (matter) and five sense-perceptions is loosened. Both Buddhism and Vedanta point to the highest level of pure consciousness, 'enlightenment', whereby the practitioner's consciousness detaches from the body and enters Nirvana. The metaphysics of consciousness therefore demands a transcendental theory that overcomes the pitfalls of scientific empiricism, reductionism, and naturalism that denies the existence of Nirvana. This demand is underpinned by findings regarding the impairment of self-referential processes typically found in long-term meditators who practise the training of their attention away from self-reference and pursue the goal of attaining a peaceful state of mind described in such terms as "bliss" and "all oneness".

Psychedelic drugs/medicines can also induce ego dissolution and spiritual experiences. The common features of which include feelings of profound joy and peace, and a sense of oneness with the universe. Experiments with psychedelics, such as psilocybin, LSD, and ayahuasca, show that the 'psychedelic state' is characterized by substantially decreased levels of activity and functional connectivity in the default mode network (DMN) area of the brain. Scientific research shows that under the influence of psychedelics, no increases in oscillatory power are observed in *any* brain region, which suggests that the brain does not produce spiritual experiences. Rather, it seems eminently plausible that under normal conditions the filtering mechanism of the brain is attuned, and hence restricted, to a limited spectrum of ZPF modes, while meditative practices and psychedelics remove these restrictions by detuning the filter and dissolving the ego. The brain then gains access to a wider spectrum of ZPF modes resulting in expanded consciousness and a sense of bliss and oneness. I would speculate that this is how my daughter Alice first experienced the world when her consciousness came 'on-line'. The ZPF must therefore contain all knowledge, making it a quantum know-it-all or 'universal mind', which is no different from how we think of God. I would also speculate that when we die consciousness returns to the ZPF, as Ancient Greek philosopher and mathematician, Pythagoras, suggested with his metempsychosis theory.

There are many philosophical theories that suggest our consciousness does not die when we do, has been around far longer than we have, and continues indefinitely. For me, consciousness is not produced by the brain. Rather, the brain is more like a transmitter that picks up frequencies – similarly to a radio or television set. One of the arguments for assuming that the brain produces consciousness is that, if the brain is damaged, consciousness is impaired or altered. However, this doesn't invalidate the idea that the brain may be a receiver and transmitter of consciousness. A radio doesn't produce the music that comes through it, but if it is damaged, its ability to transmit the music will be impaired.

In 1996, Head of the department of anaesthesiology and psychology at the University of Arizona, Dr Stuart Hameroff, and distinguished physicist, Roger Penrose, joined forces to propose a startling theory that continues to be the subject of much debate among scientists today: human consciousness is contained in 'microtubules' in our brain cells. They contend that microtubules leave the brain and this quantum information can exist outside of the body and continue to exist after death, hence the idea of an out-of-body experience and an eternal soul. Hameroff believes that consciousness has been in the universe all along. From his intense work in anaesthesiology and research of near-death experiences, he explains that when the heart stops beating and blood stops flowing the microtubules lose their 'quantum state' but the information in the microtubules is not destroyed – it is distributed to the universe at large. If the patient is revived, the quantum information can go back into the microtubules of that patient's brain. This is where the often-repeated, vivid recollection of revived patients comes from – floating out of their bodies, or a white light at the end of a tunnel.

Needless to say, in our secular, materialistic and increasingly shallow world, the science community elites have little patience with this theory. MIT physicist Max Tegmark published a critique where he argued that the quantum states of microtubules would, after death, survive for only ten seconds, and absolutely cannot have an out-of-body existence. In reply, Hameroff and others argued that microtubules could be shielded against the limited, physical environment of the brain when life expires. They faulted Tegmark's analysis for using not their original criteria, but entirely different criteria in his testing of their quantum theory, thus changing the assumptions behind their theory. Many other scientists have joined this argument, and as recently as 2014, respected physicist Anirban Bandyopadhyay confirmed from his own research that the shielding of microtubules from a dying brain environment does indeed occur.

Hameroff and Penrose have therefore given us a thought-provoking theory of how the smallest types of matter might make up the very foundation of what we call consciousness, the soul, and even ghosts. Let's face it, who wouldn't want to believe that their dead loved-ones have continued on in the aether and reincarnated into another human entity? When drawing conclusions, we must also consider a compelling observation by a reputable mathematical physicist and author, Dr Henry P. Stapp, who backs the Hameroff/Penrose position and goes one step further by submitting that a person's personality can exist as a mental entity after death. Outlandishly, he suggests that if these entities can manage to pull themselves back into the physical world, things like channelling and possession by mediums can actually happen. If this sounds too far-fetched, consider that the human brain is essentially the most complex object in the known universe. This three-pound mass of buttery consistency is an enigma in that it is 'us', and yet seems to be an utterly alien system. Despite its unimaginable complexity and power, approximately 99% of externally derived sensory data is immediately discarded upon entering the brain. The massive amount of computation required to process all of the information inundating our senses at any given time is unnecessary and beyond what the brain is capable of processing. However, I would suggest it is perfectly plausible that 'evolved' humans are capable of seemingly 'supernatural' feats as they gain more access to the untapped capabilities of the brain.

There are many more scientific theories of consciousness, and one that is definitely worth mentioning – before I move on to what religion and spirituality have to say about this most mystical of subjects – is Integrated Information Theory (IIT). Neuroscientist Christof Koch lays out this theory in his book, *The Feeling of Life Itself: Why Consciousness Is Widespread But Can't Be Computed*, proposing that consciousness is not a binary, on-or-off phenomenon. Rather, by analysing the structure of a system's circuitry, it should be possible to derive a numerical measurement of the system's cause-and-effect power, known as phi (that is, the Greek letter Φ). If phi is zero, then the system doesn't exist for itself. A system that has phi that is positive, say five, ten or twenty million phi, exists for itself. The larger this number, the more conscious the organism is. Looking beyond humans, animal brains show a lesser degree of cause-and-effect, but the effect is still there. Koch has come around to the view that all forms of life – from apes, dogs and dolphins all the way down to microbes – possess a modicum of consciousness. Another implication of IIT is that it would be impossible for computers to become conscious, at least if they are merely advanced versions of current-day machines. Computers can mimic the human brain, including mimicking the program in the brain for waking up and saying, "I am conscious," but it's all going to be the ultimate deep-fake. It will not 'feel like' anything to be a seemingly-conscious machine, which is why one might argue that humanity needs religion and spirituality for moral guidance when it comes to creating these machines and how we will treat them. God forbid it will be any worse than we currently treat animals and each other.

Chapter 2

Religion & Spirituality

I've always been fascinated by religion; supernatural stories written by men attempting to explain life's deepest mysteries. Religions usually fade and become extinct because they are the works of people. Although religion and spirituality are related, they are not synonymous. Religion is the creation of identity and culture whereas spirituality is the direct personal relationship with a Higher Power. Religions often degenerate into convention, ritual, and corruption because they are imperfect. When their creators fade, even the holiest words gradually lose their power. I think of religions in the same way Israeli intellectual, historian and professor, Yuval Noah Harari, wrote about them in his book *21 Lessons for the 21st Century*: "Throughout history, humans have created hundreds of different religions and sects. A handful of them – Christianity, Islam, Hinduism, Confucianism and Buddhism – influenced billions of people (not always for the best)." Through Twelve Step recovery, however, I began to understand that religion can be a necessary stepping stone to truth. Often, people that have 'awakened', like I had, need to explore religion before they can arrive at the ultimate truth – that there is only one underlying reality. As the great mystics, prophets and shamans consistently state: All Is One.

Christians would say that to be awakened, to be conscious, means to align one's consciousness with 'Christ Consciousness'. I am not a Christian but I became aware that consciousness is within all of us in 2013 – after I began practising a Christian meditation. I realized then that the teachings of Jesus Christ were not just words of wisdom but seeds of pure consciousness that I already knew inherently and carried inside me. Through the process of receiving these teachings by reading the *King James Bible* and the continuation of my meditation practice, Christ Consciousness simply grew and developed from within, and I had a number of massive 'Christian-esque' spiritual experiences that I wrote about in my second book *Anonymous God*. At first, I was startled by these experiences, so radically different from everything I knew before, but after a while it all started to make perfect sense. To be awakened, to be fully conscious – and aware – means to be aligned with Christ Consciousness, as one fully accepts and embraces all of Christ's Principles: love, compassion, patience, forgiveness, generosity, peace, faith, divinity, charity, non-judgment, humility, gratitude, and oneness.

The *Bible* is clearly the word of man and not the word of God, but it does contain much wisdom including the notion that we are all equal and we are not separate in any way from our own divinity. If everyone could recognise this then we would truly witness great positive changes on a global scale. In order to contribute to the creation of a more balanced society, including more respect towards 'Mother Earth', we must first put our energy into our own awakening. The answer to human evolution, therefore, lays in the shift of consciousness that must occur when we are called upon to open up spiritually and be vulnerable in our humanity, thus kickstarting the process of enlightenment. This is the biggest difference between being in a 'sleeping state' or an 'awakened state': the level of personal development one has achieved that propels us to live in an open-hearted state instead of a closed-hearted state. The key is to be found not just in reading and understanding the teachings of enlightened 'masters' such as Jesus, but in being able to wisely embody the teachings, so we can open ourselves up, regain sovereignty upon our consciousness, and gift peace, love and wisdom to others in order to help raise the consciousness of society.

The concept of a 'Higher Consciousness' is pervasive in

all religions. 'Taqwa', for example, is an Islamic term, which is used in the Qur'an over 100 times and has many definitions including: 'God-consciousness', the 'experience of awe of God', and a 'high state of heart, which keeps one conscious of Allah's presence and His Knowledge'. Taqwa motivates the person who possesses it to perform righteous deeds and avoid forbidden activities. In Sufism (Islamic mysticism), taqwa has several degrees regarding 'awakening' and awareness. The highest ranked are those who distance themselves from everything that separates them from God. One of the main goals in Sufism is to get closer to God, as the state of being separate from Allah is a misery equivalent to living in the fires and torments of hell according to Sufis.

To be truly 'awake' is simply to become aware that the highest consciousness is the consciousness of God within us; the part of a human that is capable of transcending animal instincts. This concept is also a central tenet in popular contemporary spirituality that was significantly developed by German Idealists in the late 18th and early 19th centuries – but actually has roots dating back to India's ancient Hindu holy books the *Bhagavad Gita* and the *Vedas*. The earliest historical mention of consciousness was written in Sanskrit around 800 BCE in the *Upanishads*, which are the mystical and philosophical teachings of the *Vedas*.

Vijñāna (Sanskrit) or viññāṇa (Pāli) is translated as 'consciousness', 'life force', 'mind', or 'discernment' in Buddhist literature, describing the mental force that animates the otherwise apathetic material body. Siddhartha Gotama, the Buddha, illustrated consciousness in the following manner: "It cognizes; thus it is called consciousness. What does it cognize? It cognizes what is sour, bitter, pungent, sweet, alkaline, non-alkaline, salty, and unsalty. Because it cognizes, it is called consciousness." In early Buddhist texts consciousness is discussed in at least three different but related contexts:

1. As a result of the five senses: sight, smell, hearing, touch and taste.
2. As one of the five aggregates of clinging at the root of all suffering (dukkha).
3. As one of the twelve causes of 'Dependent Origination', which the Buddha used as a template for notions of karma, rebirth and release.

Dependent origination (also known as Dependent Arising) is a key doctrine of the Buddha's philosophy, which states that all phenomena (dharmas) arise in dependence upon other dharmas: "If this exists, that exists; if this ceases to exist, that also ceases to exist," said the Buddha. Buddhism asserts that there is nothing independent, except Nirvana, which is said to be a state of emptiness whereby desire, aversion and ignorance are extinguished. The Buddhist belief that causality is the basis of metaphysical 'being', not a creator God, nor the ontological Vedic concept of a universal-self known as Brahman, nor any other transcendent creative principle, does not answer the question of 'first cause'. Yet the existence of Nirvana, confirmed by the Buddha, sounds akin to a universal background field (ZPF), as described in the introduction, from which everything manifests and returns.

Similarly, the Chinese philosophy of Taoism (or Daoism) states that the Tao (or Dao) is the force or 'the Way' that brings everything into existence. Those that follow Tao declare that gods do not exist and they seek a relationship with the divine in which there is no division. They seek a state of oneness. According to Chinese-American philosopher, Deng Ming-Dao, author of 365 Tao: Daily Meditations, "The task of following Tao is to cease all distinctions between self and the outside world. It is only a matter of convenience that we label things inside and outside, subjective and objective. Indeed, it is only at elementary stages that we should talk of Tao to follow. For true enlightenment is the realization that there is a Tao to follow but that we ourselves are Tao. That understanding comes from a simple breaking down of a wall, a shattering of the mistaken notion that there is something inherent in this life that divides us from Tao. Once the wall is broken, we are inundated by Tao. We are Tao... When you can bring yourself to the understanding that there is no difference between you and Tao and that there is no difference between meditation and 'ordinary' activities, then you are well on your way to being one with Tao."

In my opinion, one can simply replace the word God with Tao and Tao with consciousness, as they are different labels for exactly the same concept, which again reinforces the scientific idea of a ubiquitous source-field (the ZPF). It stands to reason, therefore, if a person becomes one with Tao/God/Nirvana/ consciousness then there is no division between them. They are Tao/God/Nirvana/consciousness and Tao/God/Nirvana/ consciousness is them. This doesn't mean they can do all the things that gods are supposedly able to do, but they attain a state of being and understanding where there are no distinctions, fears or uncertainties about what is divine.

It is certainly fair to say that man has been enabled, in every age and every country, to perceive things in the interior or invisible world; the spiritual dimension. The 19th century Transcendentalists saw the entire physical world as a representation of a higher spiritual world. They believed that humans could elevate themselves above their animal instincts and partake in the spiritual dimension by acquiring 'Theosophia' (God-knowledge), thus attaining a Higher Consciousness, which carried the mind from the world of form into that of formless spirit. Transcendentalism is a philosophical movement that developed in the late 1820s and 1830s in the eastern United States. It arose as a reaction to protest against the general state of intellectualism and spirituality at the time. Transcendentalism emphasizes subjective intuition over objective empiricism.

Adherents believe that individuals are capable of generating completely original insights with little attention and deference to past masters. The conservation of an undisturbed natural world is a core belief of Transcendentalism – resulting in an inherent scepticism of capitalism, westward expansion, and industrialization. Another core belief of Transcendentalism is in the inherent goodness of all people and nature, and while society and its institutions have corrupted the purity of the individual, people are at their best when they are truly self-reliant and independent. It is only from such 'real' individuals that true community can form according to Transcendentalists. Even with this necessary individuality, they also believe that all people are outlets for the 'Over-Soul' because the Over-Soul is One – uniting all people as one being: All Is One.

Transcendentalism emerged from the scepticism of Scottish philosopher David Hume, plus German Romanticism, transcendental philosophy, German Idealism, and the Biblical criticism of German philosophers, such as Johann Gottfried Herder, Friedrich Schleiermacher and Immanuel Kant. Major figures in the transcendentalist movement included: American philosopher and poet Ralph Waldo Emerson, American philosopher and poet Henry David Thoreau, American journalist, editor, critic, and women's rights advocate Margaret Fuller, and American teacher, writer, philosopher, and reformer, Amos Bronson Alcott. Swedish pluralistic-Christian theologian, scientist, philosopher and mystic Emanuel Swedenborg (best known for his book on the afterlife, *Heaven and Hell*) also had a pervasive influence on transcendentalism, which was strongly influenced by Hindu texts on philosophy and spirituality – especially the *Upanishads*.

According to the *Upanishads*, our true nature is a field of pure consciousness known as 'Spirit'. Each Upanishad, or lesson, takes up a theme ranging from the attainment of spiritual bliss – to karma and rebirth – and collectively they are meditations on life, death and immortality. The essence of their teachings is that truth can be attained by faith rather than by thought, and that the spirit of God (the Brahman) is within each of us. The *Upanishads* describe the relationship between the Brahman and the Atman. The Brahman is the universal self or the ultimate singular reality. The Atman is the individual's inner self, the soul. A central tenet of the *Upanishads* is 'tat tvam asi', which means the Brahman and the Atman are identical. There is only one universal self, and we are all one with it. The *Isha Upanishad* states, "the Brahman forms everything that is living or non-living... the wise man knows that all beings are identical with his self, and his self is the self of all beings."

The Holy Spirit may be the nearest translation of the Brahman in Christian language. The Brahman is God, the universe, and in His transcendence and immanence He is also the Spirit of man, the Self; the Atman. Thus, the momentous statement is made in the *Upanishads* that God must not be sought as something far away or separate from us but rather as the innermost part of us; our Higher Self above the limitations of our little self. When the sage of the *Upanishads* is pressed for a definition of God, he remains silent, meaning that God is silence. When pressed for a positive explanation of God he utters the sublimely simple words, "Thou art That," meaning God is everything including you. Of the *Upanishads*, German philosopher Arthur Schopenhauer said, "In the whole world there is no study so beneficial and so elevating as that of the *Upanishads*. They are products of the highest wisdom. They are destined sooner or later to become the faith of the people."

I agree that the *Upanishads* encapsulate the highest form of religion, as they insist on morality, meditation, and the worship of God in Spirit (consciousness), and they are not encumbered by such traditional dogmas and miracles that are seen as the hang-up of many other religions. Their central principle, that there is one supreme reality that manifests Itself as the universe, is not asserted as dogma. It is the ultimate truth at which it is possible for human understanding to arrive. The progress of science and spirituality does not conflict with it, in fact they only seem to confirm it. Objective reality does not exist according to quantum physics, which is exactly what the *Upanishads* declared thousands of years ago. The *Upanishads* and quantum physics both say that the observer and the observed are the same thing. In his 1944 book, *What is Life?*, Austrian-Irish physicist, Erwin Schrödinger, took on a peculiar line of thought: if the world is indeed created by our act of observation, there should be billions of such worlds, one for each of us. How come your world and my world are the same? If something happens in my world, does it happen in your world too? What causes all these worlds to synchronise with each other? He found his answer in the *Upanishads* and wrote: "There is obviously only one alternative, namely the unification of minds or consciousnesses. Their multiplicity is only apparent, in truth there is only one mind. This is the doctrine of the *Upanishads*."

According to the *Upanishads*, everything we see around us is 'Maya': a distortion of the Brahman caused by our ignorance and imperfect senses. The *Chandogya Upanishad* says, "All this is Brahman. Everything comes from Brahman, everything goes back to Brahman, and everything is sustained by Brahman." Quantum physics backs up this ancient wisdom confirming that reality exists as waves, and wave-particle duality arises due to our observation. Because we cannot perceive the true wave nature of reality, our observation reduces it to the incomplete reality we see. This reduction is what we know as the collapse of the wave function. Thus, the concept of Maya neatly maps to this collapse.

One of the key messages of the *Upanishads* is that the Spirit (the Atman) can only be known through union with God, and not through mere learning. Not through discourse, not through the intellect, not even through study of the scriptures can the

Self (Spirit) be realized. The Self only reveals Himself to the one who longs for the Self. Those who long for the Self with all their heart are chosen by the Self as his own. Freed from the fetters of separateness, they attain immortality, meaning they no longer fear death, as Spirit is eternal. Thanks to my spiritual experiences with Vipassana meditation and psychedelics, I too no longer fear death. I couldn't sum up my understanding of death any more coherently than one of my heroes, legendary Hip-Hop artist and leader of the Wu-Tang Clan, the RZA, who wrote in his wonderful book, *The Tao of Wu*: "Infinity is a natural plane of our energy. That's where you are super-conscious. I think you can, at that moment (of death), make a choice. At the end of your body's physical life, when your physical animated energy's running out, you'll know it. This wisdom is also reflected in the Tibetan teachings that guide you through that moment of consciousness between death and rebirth."

The *Upanishads* describe how reality arises out of consciousness, but consciousness cannot be found inside our bodies as a substance or an organ. How can a non-material consciousness, therefore, interact with – and control – our material bodies? Exactly where does mind interact with matter? As already discussed, this mind-body dualism problem has perplexed scientists and philosophers alike for a long time. Since we haven't been able to locate or explain this interaction, we're left with a deceptively simple choice: either consciousness doesn't exist, or reality doesn't.

The answer therefore is: there is no objective reality that exists independently of the observer. In other words, if you didn't exist in the form that you currently exist, the universe would look, smell and feel nothing like how you experience it in your human form. This idea has little appeal to Western thought because it is dubbed "unscientific". But it is unscientific only because our science is based on objectivation, whereby it has cut itself off from an adequate understanding of consciousness.

The material world as we understand it scientifically has been constructed at the price of taking the Self (consciousness) out of it, even labelling it an 'illusion'. However, my experiences with the psychedelic medicine ayahuasca (in 2017) and psilocybin mushrooms (in 2019) proved to me beyond any reasonable doubt that the material universe, as we experience it, is manifesting from a ubiquitous background field of energy-consciousness that can only be accessed when the Default Mode Network of the brain is taken off-line.

Chapter 3

Psychedelics

The idea that the Old Testament prophets may have been using psychoactive substances in order to evoke a shamanic trance – in which the revelations of Yahweh could be received – is as troubling for modern day Jewish and Christian believers as Darwin's Theory of Evolution was to their 19th century counterparts. Just as Darwin's theory of evolution challenged the myth of creation from the *Bible's Book of Genesis*, the proposed entheogenic origin for these religions offers a scientific and anthropological theory based on shamanism and psychoactive plants. As Swiss classicist Professor Georg Luck has noted: "The idea that Moses himself, and the priests who succeeded him, relied on 'chemical aids' in order to touch with the Lord must be disturbing or repugnant to many. It seems to degrade religion – any religion – when one associates it with shamanic practices." Luck experienced these reactions himself when his decades of research into magic rites in the ancient world drew him to such a hypothesis. "As I was doing research on psychoactive substances used in magic and religion in antiquity, I happened to come across chapter 30 in *The Book of Exodus* where Moses prescribes the composition of sacred incense and anointing oil. It occurred to me, judging from the ingredients, that these substances might act as 'entheogens', the incense more powerful than the oil."

Psychedelic medicines can be used to alter brain cognition and perception. Many people, including myself, believe this to be a higher state of consciousness, as these altered states have been known to result in a long-term and positive transformation of the person ingesting the medicine. Psychedelics can also be used for psychoanalytic therapy as a means of gaining access to higher levels of consciousness and can provide patients the ability to access memories that are held deep within their unconscious mind. The word psychedelic essentially means revealing what is hidden, which is why the use of psychedelics, in the appropriate set and setting, can reveal the symptoms of illnesses buried in our subconscious. Due to the insights that psychedelics can produce, I've heard it said that they are analogous to ten years of psychotherapy in one session, and like 'cheat codes' for the game of life. This has certainly been my experience.

The term psychedelic is derived from the Greek words 'psyche' (soul/mind) and 'delein' (to manifest), hence, 'soul-manifesting'. The implication being that psychedelics can access the soul and develop unused potential of the mind. The word psychedelic was coined in 1956 by British psychiatrist Humphry Osmond, and championed by American psychologists Timothy Leary and Richard Alpert (later to become known as contemporary spiritual teacher Ram Dass). Many psychedelic drugs are illegal worldwide under the UN conventions unless used in a medical or religious context. Despite these regulations, the traditional use of psychedelics has always been common across the world, as they have a long history of medicinal and religious use and they are prized for their perceived ability to promote physical and mental healing. In this context, any plant or drug taken to facilitate spiritual experiences are known as 'entheogens', which translates to 'God within'. Native American practitioners using mescaline-containing cacti, such as peyote, have reported success in eradicating alcoholism and other addictions, and Mazatec shamans routinely use salvia divinorum (also known as 'sage of the diviners') and psilocybin 'magic' mushrooms for divination and healing. Ayahuasca is used in Peru and other parts of South America in religious festivals as a physical and spiritual remedy. In the 1960s, at Harvard University, Leary and Alpert researched the possible therapeutic applications of psychedelic drugs, such as psilocybin and lysergic acid diethylamide (LSD), before they were shut down by President Richard Nixon's 'War on Drugs' (a global campaign to reduce the illegal drug trade led by the U.S. federal government, which is still raging today).

Alcoholics Anonymous Co-Founder, Bill Wilson, believed that LSD might be used to cure alcoholics, and credited the drug with helping his own recovery from often debilitating depression. About twenty years after setting up the Ohio-based sobriety movement with Dr Bob Smith in 1935, Wilson came to believe that LSD could help "cynical alcoholics" achieve a spiritual awakening and start on the path to recovery. Wilson initially thought that LSD could help people to understand the alcohol-induced hallucinations experienced by alcoholics. He thought it might terrify drinkers into changing their ways, but after his first 'acid trip' in 1956, Wilson began to believe it was insight, not terror, that could help alcoholics recover. He believed LSD, by mimicking insanity, might help alcoholics achieve a central tenet of the Twelve Step Program; finding a Power greater than themselves that could restore them to sanity. About Bill Wilson's experience, author Aldous Huxley wrote to priest Father Thomas Merton, in 1959, "A friend of mine, saved from alcoholism, during the last fatal phases of the disease, by a spontaneous theophany, which changed his life as completely as St. Paul's was changed on the road to Damascus, has taken lysergic acid two or three times and affirms that his experience under the drug is identical with the spontaneous experience which changed his life – the only difference being that the spontaneous experience did not last so long as the chemically induced one. There is, obviously, a field here for serious and reverent experimentation."

Unfortunately, experimentation with LSD and other psychedelics was cut short due to the War on Drugs, but thankfully there has been a resurgence in the research in recent years. Thanks to Wilson's experimentation, psychedelic medicines became part of my recovery journey in September 2017 when I attended an ayahuasca retreat in Mallorca. I had to do three years of research on psychedelics – plus an eight-hour psychedelic trip with ayahuasca – before it was revealed to me that the material world, as we perceive it, emerges from consciousness. Thanks to my research, I knew that success (in the sense of processing my trauma and healing from it) hinged on integration after the ayahuasca ceremony, and my intentions before the ceremony, which were: to figure out if my ex-partner (at the time) Adele and I were meant to be together; to drop selfishness; to become more open and honest in my relationships; and to meet God. Thankfully, I did meet my Higher Power, Adele and I are now married and have a daughter together, I am rarely selfish these days, and I am far more open and honest in my relationships. My experience with psilocybin in 2019 was the final part of my integration of these lessons, as I fully understood that life is not happening to me, but rather it is happening for me – to learn and grow as a human.

Although any drug, including psychedelics, can be used recreationally and as a means of escape, mushrooms (and other plant medicines such as ayahuasca, peyote, and salvia divinorum) are not addictive, and when used in the appropriate set (mindset) and setting (the conditions under which you 'trip') they can have hugely therapeutic benefits. This quote from American mycologist and author Paul Stamets embellishes my point: "Mushrooms are miniature pharmaceutical factories, and of the thousands of mushroom species in nature, our ancestors and modern scientists have identified several dozen that have a unique combination of talents that improve our health." This can be said of all psychedelic plant medicines.

Research into ayahuasca, which contains the powerful psychedelic N,N-Dimethyltryptamine (DMT), has shown promise for alleviating some mental illnesses and for providing other long-term health and social benefits among regular drinkers of the brew (in ritualised and religious community contexts). The ritual use of ayahuasca doesn't typically produce health or psychosocial problems, such as addiction, and has actually been correlated with lowering the amount or severity of substance dependence. Research found that among a randomly selected group of Brazilian ayahuasca church members, the majority reported prior history of moderate to severe problems with alcohol or other drugs. All had stopped using substances (including tobacco), other than ayahuasca, after joining the church – and attributed their improved health behaviours to drinking ayahuasca. The subjects also reported less excitability and impulsivity, and more confidence and optimism compared with matched-control community members who didn't use the medicine.

Ayahuasca expands one's consciousness beyond our five-sensory default reality, so one can experience 'more'. It supposedly opens doors to truths about past, present, and even future stories that encase suffering and long periods of confusion, enabling deep healing, clarity, and the beginning of a personal transformation. People who have consumed ayahuasca (including myself) recount having spiritual revelations regarding their purpose on Earth, and deep insights about the universe and about themselves. This is viewed by many as a spiritual awakening and is frequently described as an "unburdening" or "rebirth".

In 2016 I felt a calling from deep within to 'journey' with ayahuasca. Initially I ignored this calling until it became clear to me that it was something I needed to do, as I felt I couldn't keep writing about the subject without experiencing it for myself. My experience was beautiful and profound. The overall event can be summed up by saying that I slipped beyond the confines of my ego and was enlightened to the fact that I am interconnected with all sentient beings, and connected to the Élan Vital (the universal Life-Force that we most commonly refer to as 'God' or a 'Higher Power'), as proposed by French philosopher Henri Bergson in his 1907 book *Creative Evolution*.

It is written in *The Tibetan Book of The Dead*: "Remember the clear light, the pure clear white light, from which everything in the universe comes, to which everything in the universe returns; the original nature of your own mind. The natural state of the universe unmanifest. Let go into the clear light, trust it, merge with it. It is your own true nature, it is home." This perfectly captures my profound, life-changing experience under the influence of ayahuasca, in which my understanding of my 'Self' as a physical entity completely disbanded – and what remained merged with the pure white light as I became one with 'Source'. I was 'home' in that womb-like place for what felt like an eternity but was actually no more than a couple of hours, but when I emerged from the bliss of 'Samadhi' my knowledge and understanding of consciousness had evolved.

Psychedelics, for years labelled as illegal, taboo, recreational drugs, with no scientific or medical value, are now being rediscovered for their extraordinary therapeutic potential. Psychedelic researchers are increasingly being welcomed back into the fold of large institutional structures that had for years ostracized this kind of study. It is relatively easy for previously close-minded scientific communities to understand modern psychedelic research when it is focusing on a drug's therapeutic value. A subjective psychedelic experience may be somewhat eccentric and obtuse, but if we can slot it into a clinical trial structure and show it to be effective in treating specific conditions, then we can legitimize it. While psychedelic science itself still lives on the outskirts of mainstream science, the world of psychedelic science has its own fringes. This research is not explicitly focused on specific therapeutic outcomes, but instead on investigating subjects as broad as the very foundations of consciousness.

In the 1990s, author of *DMT: The Spirit Molecule*, and former Professor of Psychiatry at the University of New Mexico, Dr Rick Strassman, conducted a five-year long study of DMT. In 2019, Johns Hopkins University launched the first dedicated psychedelic research centre in the United States: the Center for Psychedelic and Consciousness Research. This multimillion-dollar enterprise is primarily focused on studying the therapeutic potential of psilocybin, from its uses in depression and anorexia, to its potential as a smoking cessation tool, but that is not all that is going on there. It is no accident that the name of the centre mentions psychedelic and consciousness research, as Professor of Psychiatry and Neuroscience, Roland Griffiths, the centre's director, is a true pioneer of both psychedelic research and consciousness. In the early 2000s, Griffiths led a landmark study investigating how high-dose psilocybin can produce mystical experiences of religious or spiritual significance. This was the first study using psilocybin in decades to receive regulatory approval by the Federal Drug Administration (FDA) in the United States. Some of their work investigated the weirder edges of psychedelic science, and the latest study published by the team explored the variety of experiences people have when encountering strange autonomous entities after smoking DMT.

The specificity of this research grew out of a survey by the team exploring what were deemed "God encounter experiences", triggered by classical psychedelic substances. That study laid the foundation for a new and much more targeted investigation. There is a significant volume of subjective anecdotal accounts describing highly detailed encounters with strange entities while under the influence of DMT. A 2006 report, which compiled 340 detailed DMT 'trip' reports, found 66% of experiences recounted interacting with independently existing entities while under the influence of DMT. The study gathered data from an online survey, which was accessed by over 10,000 people. After whittling down responses to only include subjects with no prior diagnosis of psychosis, who were reporting pure DMT experiences that led to encountering a specific autonomous entity, the study was left with over 2,500 quite surprising responses. The breadth of experiences reported in the survey differed from the classically iconic narratives chronicled in Rick Strassman's landmark work. In fact, it really surprised the team how many people were describing experiences that were quite different from what Rick Strassman's subjects described, which suggested there is a much larger variety of experiences that people have. "Being" or "Guide" were the most common descriptive terms used to label the entity people encountered during a DMT experience. The vast majority of respondents described benevolent, loving, joyful spirits or beings that were there to communicate and relay information in some way.

Gauging the metaphysical implications of these encounters, a striking 80% of all subjects reported the experience as altering their fundamental conception of reality. The majority of respondents labelled the experience as more real than their everyday waking life, and 72% claimed the entity encountered continued to exist after the DMT experience. Perhaps one of the most intriguing findings in the study is that more than half of those subjects identifying as atheist or agnostic before the DMT experience no longer identified as such after the entity encounter. This type of decrease in atheistic self-identification following psychedelic experiences is not a particularly novel finding. Prior research into psychedelic-induced spiritual experiences has seen similar results. However, it is remarkable that such a single, short experience, lasting no more than half an hour in most cases, can generate such powerful changes in subjective belief. What was seen in a variety of reports was an indication that the experience was so deeply profound that it was difficult to not have it shatter one's preconceived belief system. They went from a firm belief in nothing, to potentially a belief in something. Their experiences were so visceral and real that it couldn't not change their understanding of their place in the universe. The vast majority of the 2,500 reports classified the DMT-induced entity encounter as one of their "most meaningful, spiritual, and psychologically insightful lifetime experiences, with persisting positive changes in life satisfaction, purpose and meaning attributed to the experience".

Although I'm an advocate for using psychedelics in the appropriate set and setting to explore consciousness, before I got sober I had many hallucinogenic experiences on a cocktail of synthesized drugs including alcohol, cocaine, ketamine, and MDMA. During one such experience, I was driving my car down a dual carriageway and the car morphed into a boat, the road morphed into the sea – and the roundabout ahead of me became a humpback whale. I had another hallucinogenic experience at the Global Gathering dance music festival whereby lightning came down from the roof inside the tent. These experiences, and many more, ultimately led me to my rock-bottom and recovery, so I do not discount them, however, I do not consider them part of my spiritual development. I would be mortified if my daughter grew up to take drugs in the same irresponsible and reckless way that I did. Unlike psychedelics, which I see as consciousness expanding medicines that (when taken in the right set and setting) can be useful technologies to facilitate spiritual experiences, I do not feel that drugs (including alcohol) can be used in the same way – and I would never advocate their use.

Chapter 4

Drugs (including alcohol)

One of the most important observations I made in my first book, *Addiction Prevention: Twelve Steps to Spiritual Awakening*, is that alcohol is no different from any other drug in the sense that it is a psychoactive compound with sleep-inducing and euphoric properties as well as being addictive due to its effect. The ingestion of alcohol results in a variety of changes to consciousness. At rather low doses alcohol is associated with feelings of euphoria but as the dose increases, people report feeling sedated. Generally, alcohol is associated with decreased reaction time and visual acuity, lowered levels of alertness, and reduction in behavioural control. With excessive alcohol use, a person might experience a complete loss of consciousness and/ or difficulty remembering events that occurred during a period of intoxication, which I did many times over the years!

Ironically, I took my first alcoholic drink the same year my dad stopped drinking and started his recovery journey but I opted not to continue as I was a keen sportsman and I knew what had become of famous alcoholic athletes such as Manchester United legend George Best. I promised myself I'd never end up an alcoholic like my dad, but in hindsight, I had little choice in the matter because I have a physical intolerance to alcohol, which I was either born with or developed over time. With schoolmates, I became known for not drinking, which was uncommon, until I was 17 when I drank some alcoholic beverages at a soccer awards ceremony – where I gained accolades for my accomplished performances. This time the alcohol had a profound effect on me. I thought I was drinking to be sociable but in retrospect, it was because I was no longer resilient enough to resist peer group pressure, and I didn't have the courage to rebel against the status quo anymore, so instead I copied my friends – and I got the 'taste' for it. There was a real sense of ease and comfort that came with those first few drinks – an inner warmth as if my pilot light had been switched on – like I was lit up from the inside. I instantly became funnier, better looking, more confident, and more intelligent – or so I thought! From that point on, my obsession for sport was replaced by an obsession for alcohol, music, drugs and sex – usually in that order!

When I left home for university in 1999 it felt like the 'shackles' were off and I hastily set off down the road to freedom, but I took an accidental wrong turn down a path to the destination marked 'self-destruction'. Blackouts, whereby my conscious mind switched off but my body stayed out to play, morning vomiting, and plenty of guilt and shame following drunken shenanigans became a regular occurrence. I perceived that I was getting so much enjoyment at the loss of my inhibitions that I was to drink alcoholically for the next ten years. In retrospect, alcohol never made me happy. It merely rendered me unconscious, oblivious to my feelings and my surroundings, and gave me a false sense of pleasure.

At university I 'partied' almost every night, and I didn't learn much else other than how to drink and take drugs. Practically everybody I hung-out with during my time at university used drugs (including alcohol) like I did, but as we grew older, they either calmed down, stopped altogether, died in various ways, or like me – they carried on. Throughout my early twenties my anxiety and depression worsened progressively as time went by, but nobody could have known that I was anxious and depressed because I never spoke about the way I felt. From my mid-to-late twenties my substance and behavioural addictions were like a game of 'whack-a-mole'. Every time I 'whacked' one mole down another mole popped up in the form of alcohol, sugar, caffeine, sex, shopping, social media or drugs; whatever changed the way I felt – until it stopped working, then I would find a new 'fix'.

I DJ'd and danced for over ten years in the underground hard house music scene, which was insanely (and in hindsight tragically) fun when I was in my early twenties. Everyone has a dark side and mine is inextricably linked to electronic drug-music. I frequented debauched places, such as Hard house, trance and techno clubs, plus the inevitable afterparties that followed. My mind was becoming warped the more I poisoned it with drugs. Hard house was a very niche music to be into at the time. It was popular on the underground dance music scene from 1998 and reached its peak in 2003 when BK's track *Revolution* received regular plays on Radio 1. But from that point on its popularity began to wean and most clubbers and DJs began to branch off toward the House, Techno, and Trance scenes. The release of *Hostile* by Paul Glazby in 2003 was, I believe, the death siren for Hard house music, as it was simply too hard and too dark and the scene had lost its appeal to the gay and female cohort of clubbers – who made the scene so uplifting, vibrant and energetic. I remember the first time I heard *Hostile* played by Paul Glazby himself at Base @ Space in Leeds and thinking to myself, "Wow, this is mega-hard! Where do we go from here?" It was the perfect metaphor for my own addiction. Where was I to go from there? The only way was down.

I didn't stop raving until 2013, long after the Hard house party had ended. Certain Hard house brands, such as Trade, Tidy, Storm, Sundissential and Parlez-vous? are almost the stuff of legend and have cult-like status among veteran ravers like myself. There was a mystique about the Hard house scene and being part of it as both a DJ and a clubber meant I felt part of a tribe (something I didn't feel again until I joined Twelve Step fellowships). Couple this with copious amounts of drugs and it's no wonder my brain has such a fondness for the music, as the 'good times' have been etched into my neural pathways forever. I only have to hear a few bars of my favourite Hard house tunes and the euphoric recall catapults my consciousness back to those heady nights in the early millennium.

By 2006 I had transitioned from Hard house to House and Techno music but my love for Hard house has never died, and if I had to pick my favourite genre of music, that would be it. Hence why it has been so difficult to let go of my association with the scene. I put down my headphones in 2008 before picking them back up again in 2011 in an attempt to achieve my goals, which I did. As one half of the Lox & Leigh Green DJ/ Producer duo I was signed to Most Wanted DJ Management and DJ'd in Ibiza with the rest of the Most Wanted DJ roster and in the main room at the Ideal Weekender. This was a dream come true but at the time it felt empty because by that point I had awakened and realized that fame is empty, success is empty, casual sex is empty, and the only things that have any real substance for me (and bring me happiness) are creativity for the sake of creativity, and unconditional love.

I know today that happiness is something that comes from within and if you are truly happy nobody can affect your happiness or take it from you. Happiness is not to be attained in material pleasures, and the only way to achieve true happiness is through unconditional love for other people through selfless service. Once you attain love by helping others, you experience peace, and finally you attain happiness, which is total and complete satisfaction with yourself. Happiness therefore arises out of a peaceful and contented mind. When you are peaceful and content, you feel whole; you can enjoy simple, blissful moments, such as observing the sunset, or cuddling your child because your mind isn't wandering off thinking how to fill the void with material pleasures. Indian-American entrepreneur, Naval Ravikant, explained happiness in a 2020 Facebook post in a way that really resonated with me: "I struggled for a lot of my life to have certain material and social successes, and when I achieved those material and social successes, or at least beyond a point where they didn't matter as much to me anymore, I realized that my peer group, and the people around me, the people who had achieved similar successes, and were on their way to achieving more and more successes, just didn't seem all that happy. In my case, there was definitely hedonic adaptation. I'd very quickly get used to anything. What led me to the conclusion, which seems trite, is that happiness is internal. That set me on a path of starting to work more on my internal self and realizing that all real success is internal and has very little to do with external circumstances. One has to do the external thing anyway. That's how you're biologically hard-wired. It's glib to say, 'You can just turn it off.' You have to do it, and you have to have your own life experience that then brings you back onto the internal path."

Happiness is definitely an 'inside job' as we say in recovery. It doesn't matter how much 'stuff' you have, as all wise men know. The RZA summed it up nicely in *The Tao of Wu*: "Rocking too much bling can reveal a hole in a man, an emptiness he's trying to fill inside... Jewels are minerals, compressed pieces of earth, stacks of crystalline carbon. What gives them shine is their history. It's the same with man. When a man recognizes himself, he recognizes his true jewel, and his body expresses that wisdom. He becomes a jewel himself. If his mind is sharp, the way he walks and talks has a certain beauty about it. Attain wisdom and you have all the bling you'll ever need."

In psychological terms, wisdom is above the emotional and rational mind. Since I put down my headphones, then drugs (including alcohol), I've been on a spiritual journey to attain wisdom, which led to the inception of my *Life in Recovery Podcast* in 2018, and to me writing a series of books including this one. Long-form conversations with people all over the world recorded for my podcast, and the research that goes into writing my books, have enabled me to gain wisdom about Homo

Sapiens and the universe at large, which has radically changed my perception on many subjects. For example, my perception of 'rave culture' and dance-club environments changed after I got sober. I had a persistent and nagging feeling that, ethically and spiritually, DJing and the drug-fuelled 'club-scene' was no longer for me, and crucially, I was no longer interested in chasing fame, which was another addiction of mine. Hence, why, in April 2014, during the biggest gig of my DJ career at the Ideal Weekender in Prestatyn, I looked out across the sea of clubbers and made the decision to quit. I've looked back wistfully only occasionally. This isn't to say I think there is anything wrong with being a DJ. For me, a true DJ who can mix tracks seamlessly and create musical 'journeys' is an artist. It's just that the context in which DJing takes place tends to be quite 'sick' environments and I'd rather be with people who are on a spiritual journey rather than a synthesized drug-journey these days.

My own synthesized drug-journey began in 2000 at a club called Space in Leeds where I was introduced to the drug MDMA – better known as 'ecstasy' or 'Molly'. MDMA originally gained a small following among psychiatrists in the late 1970s and early 1980s, as some psychiatrists believed that it enhanced communication in patient sessions and allowed patients to achieve insights about their problems. MDMA acts on neurotransmitters in the brain to give users an extreme alteration of their mood, but can also cause a variety of unwanted cognitive defects and physical effects. When someone takes MDMA it causes a massive release of serotonin, dopamine, and norepinephrine, which can result in an enhanced sense of well-being, increased extroversion, emotional warmth, empathy toward others, enhanced sensory perception, and a willingness to discuss emotionally-charged memories. I experienced all of these benefits regularly in club environments and after-parties.

On the flip-side, MDMA can also cause a number of

acute adverse health effects including: high blood pressure (hypertension), faintness, panic attacks, and a loss of consciousness and seizures. The brain can become depleted of serotonin, which contributes to the unpleasant after-effects (known as a 'come-down') that many users experience after taking MDMA. Research has shown that the damage MDMA causes to serotonin-containing neurons can be long-lasting, and studies have found that some heavy MDMA users experience: depression, impaired attention processes, and working memory impairment. I have experienced all three as a result of my use of MDMA. A major difficulty researchers have had with evaluating the effects of MDMA on the brain is that most of the time the ecstasy tablets that users purchase on the street are not pure MDMA and contain other drugs or substances. There is also the likelihood that MDMA users are using other drugs like marijuana or alcohol (like I did), which have their own effects on the brain. Therefore, it is difficult for researchers to determine if the effects they observe are from MDMA alone, the other drugs, or a combination of the two.

One can only describe the effects of ecstasy as a synthesized spiritual experience. The first time I used it, I transcended waking consciousness as it dissolved the boundary of my ego and all my insecurities melted away. The love I felt for everyone I encountered was all encompassing and so powerful that the next day I decided I would take 'pills' as often as I could. What goes up, however, must come down, and two years later, having consumed ten 'Es' on New Year's Day and ten 'Es' the night before, I realized the ecstasy had stopped working. I was no longer experiencing the same intensity of loved-up 'high', and when I wasn't taking ecstasy, I felt utterly miserable. I was 'spiritually bankrupt', and so far-removed from my emotions that I felt empty. My mental health was suffering as I became more and more depressed – so much so that my friends intervened and told me all the reasons why I needed to stop.

Little did they know, however, that I replaced ecstasy with cocaine and for the next five years I drank alcohol every night and used cocaine regularly to help me drink more alcohol. Little did I know that cocaine can cause brain damage, even when used only a few times. Damage to brain structures can trigger addiction, which is scientifically proven to be a disease involving the brain's reward circuits and dopamine systems. Cocaine can also cause long-term damage to mental health, which appears in the form of mood or emotional disturbances. Because the drug directly interferes with dopamine being reabsorbed by neurons, one of the symptoms of a 'cocaine comedown' is serious depression. If the brain does not reach its original equilibrium then a person who has struggled with cocaine abuse for a long time may develop permanent depression and require ongoing mental health treatment. Fortunately, I only used cocaine once or twice a month because I couldn't afford any more than that, so I never became addicted to it. However, the less drugs I used, the more my drinking escalated and by 2009, at the age of 28, I needed alcohol every day and I rarely went out after work because I was indoors, on my own, drinking myself into 'oblivion'. My drinking went on like this for about a year before my girlfriend Sally suggested that we leave Leeds and go to live in Australia for a while – to get away from my problems. Of course, you take your problems with you wherever you go, as your problems are all in your mind.

"The road of excess leads to the palace of wisdom," according to English poet William Blake. Thankfully, I woke up from my drug-fuelled nightmare and set off down the path of understanding, knowledge and freedom toward the palace of wisdom. On the morning of September 7th, 2009, I awoke from (what I hope to be) my final alcoholic blackout and I knew the game was over. After overhearing Sally telling her sister she would leave me if I didn't change my behaviour, I got down on my knees and I prayed to a God I didn't even believe existed. I asked Him to help me as I couldn't continue doing what I was doing to myself and the people I loved anymore. I was sick and tired of being sick and tired. When I stood up from that prayer, something inside my chest shifted; a blockage around my heart had been dislodged and simultaneously the war in my mind (shall I drink today or not?) was over. I had surrendered to the fact I cannot drink safely. I felt total serenity – a revelation that cannot be accurately reported; I believe Catholics refer to it as being "reborn" or "Born Again".

That night I had the best sleep of my life to date. Days later, I realized the mental obsession for alcohol had been completely removed from my mind. I was surrounded by alcohol in Thailand (where we stopped off for a month), and in Australia, but I had no desire to drink. For the following nine months I didn't use any mind-altering substances, until one day, I randomly found a bag of 'Skunk' weed in the caravan park where Sally and I lived and worked. I rolled a 'joint' and smoked it on the beach. I smoked the Skunk because, after six months of mental clarity – feeling great physically and wondering why I ever used drugs in the first place – my mind descended into extremely resentful, homicidal and suicidal thoughts. It was like a washing machine, sloshing and spinning a poisonous cocktail of toxicity round and round. I constantly felt stressed and anxious and I wanted out. I thought the weed would help but it didn't. It made me feel physically sick, increased my anxiety, and I felt extremely paranoid. Cannabis consumption has historically been linked to positive feelings of euphoria and altered states of consciousness that can invoke deeper, contemplative journeys, of which I had previously experienced. But that experience was the straw that broke the camel's back for me. I wrote in my journal that evening, "I will never use drugs or drink alcohol ever again." I rang my dad the next day who suggested I try a Twelve Step recovery meeting, but I politely declined. Secretly, in my opinion, Twelve Step fellowships were religious groups and my dad was weak to need those people. I thought I could do it on my own. I was too prideful to drink alcohol but ironically too full of pride, and my own self-importance, to ask for help.

Three months later, in London, following a year in Australia, I was back down on my knees begging God for help. I broke down. I couldn't take it anymore. There was a rage inside me that I was struggling to tame, and I felt like I should drink or kill myself. I rang my dad again and told him I was going to try a Twelve Step meeting and I've been in recovery ever since with no relapses. In hindsight, drugs (including alcohol) only served to help me hit my rock-bottom and introduce me to my Higher Power: consciousness – the bedrock of my recovery.

Chapter 5

Recovery

I think every adult is in recovery to some degree: recovery from the inevitability of suffering and trauma; recovery from emotional and mental health issues; recovery from addiction. Recovery is a gradual process. It's a healing process and a spiritual process; a journey rather than a destination. There are many different paths to recovery depending on what recovery looks like for the individual. Some prefer to call it a path of discovery rather than recovery, which is fine by me. I always say it's whatever works for you that counts, and peace of mind is my personal barometer to how well I'm doing on any given day. If I don't have peace of mind then there's work that needs to be done immediately to continue my personal evolution.

When I arrived at my first Twelve Step recovery meeting in November 2010, at a church in North London, I was greeted by happy, smiley people, laughing and joking; they were very welcoming. It seemed a bit suspect. I was waiting for 'the catch'. I sat at the back of the meeting and listened, as they had suggested, for the similarities in people's stories and not the differences. There were many, especially when people spoke about their feelings, such as fear, anger, guilt, shame, remorse, and sadness. At last I knew what was wrong with me! I'd found my new tribe. After the meeting had finished, it seemed like I floated home, declaring to my fiancée Sally that I was an alcoholic and I would never drink again! It was a huge weight off my shoulders – a complete unburdening.

In my early days of recovery, people suggested that I don't make any major decisions in my first year, as it was likely I would make impulsive choices to attempt to make myself feel better now that I no longer had my perceived 'crutch' alcohol. I didn't listen, and instead, I found myself a new job, and married Sally within two months. Needless to say, the job and our marriage didn't last. I was attempting to fix my outside before fixing my inside. I had a moment of serenity on our wedding day, similar to that which I experienced the morning after my last drink, but retrospectively, deep down, I knew the relationship wasn't quite 'right' because my discreet inner-voice (consciousness) had been telling me so for a couple of years. As I began to change due to my newfound sobriety, the wheels of our relationship slowly started to come off. Couples either grow together or they grow apart, and in our case, sadly, we grew apart – mainly due to my inability to maintain an honest, loving partnership. Our nine-year relationship ended in separation, and the divorce was finalised in 2013. The break-up was one of the worst periods of my life. It upset me so much to see my wife and best friend distressed because of me, and I was so full of fear about the future that I almost drank on several occasions. I instead self-medicated, first with food, then sex with other women. It wasn't until my sponsor suggested that I "up my game" in terms of working my program – specifically praying and meditating, writing gratitude lists, taking inventory, and being of service to others – that things got better.

I left London briefly in 2014 and spent two months in Gran Canaria with my Italian friends – where I concluded that I wanted to embark on a career as a counselling psychologist. I emailed my parents to inform them that I had decided to move back to London to study. They were initially concerned about how I would pay for this education but serendipitously my mom came across an advertisement for an Apprentice Drug and Alcohol Counsellor with a London-based charity, which included training for a Level 3 Counselling diploma. The only criteria were applicants had to have at least three years abstinence from drugs and alcohol, and I was four years abstinent at the time. I sent my application from Gran Canaria, got offered an interview, moved back to London, passed the three-interview-process, and began working for the charity in a London prison in February 2015. I completed my one-year apprenticeship at the prison then transferred to another London prison for a permanent counsellor position – where I worked until 2017. During those two years I went to a minimum of four Twelve Step meetings per week and I saw three counsellors biweekly for roughly six months each, which helped me to stabilise my life and forgive myself for my past mistakes.

Thanks to following the suggestions in Twelve Step recovery and finding a Higher Power (of my own understanding), I have since acquired a sanguine view of life and changed everything to serve my life purpose, which I believe is to help people. I've experienced lots of change and I've had to make some tough choices, and ride some rough waves, but the sea has settled down and I am experiencing an ever-progressing serenity. Every now and then the waves get choppy but I know I'm safe in my recovery 'lifeboat' – providing I stay in the middle of it. I now believe that I crossed the line into alcoholism within five years of taking my first alcoholic drink. If it hadn't been for MDMA and cocaine, I feel I would have hit my rock-bottom much sooner. Cocaine enabled me to drink for much longer periods rather than 'blacking-out', and many years raving on MDMA meant I didn't drink as much alcohol as I might have done, as the two drugs don't complement each other so well.

With hindsight, it was synchronicity, "God's way of remaining anonymous," according to Theoretical Physicist Albert Einstein, that got me sober – and to where I am today. For me, synchronicity (what I used to refer to as coincidence) is an indicator that God/consciousness is making Itself known. The first time I heard that quote from Einstein it blew my mind, as a series of 'coincidences' led me into recovery and saved my life. Once I began to recover from alcoholism, my spiritual journey helped me to understand that 'coincidence' is actually synchronicity, and synchronicity is my Higher Power's main mode of communication with me. Even during the writing of this book, it has been bizarre how many times I've synchronistically accessed information that I did not previously have access to, which has bridged the gap between two sections, or enabled me to finish a section. Consciousness was pointing me in the right direction for sure. For example, I had a YouTube video saved in my inbox for months that a friend had sent to me: *Prof. B. Alan Wallace talks on "Mind, Emptiness and Quantum Physics."* Just as I was searching for information to finish the introduction to the book, I happened to be clearing out my inbox and I came across the video, which helped me perfectly.

My life story is peppered with synchronicity. I have many supernatural stories about coincidence, serendipity, synchronicity, and spiritual signposts that I wrote about in my second book *Anonymous God*, and synchronicities continue to happen since that book was published. For example, on the day my daughter Alice was born in February 2020, the Google home page depicted Alice in Wonderland, which I interpreted as an auspicious coincidence having asked my Higher Power for a sign that everything was going to be okay during (and following) my partner's C-section. It seems as though, every time I ask for a sign from my Higher Power, usually when my faith is wavering (for whatever reason), my Higher Power delivers a demonstration of coincidence, serendipity, synchronicity or a spiritual signpost to assure me It is there to support me. I have absolutely no idea how this works but it makes me think that God/consciousness is essentially a 'Universal Mind' connecting all sentient beings in the cosmos.

I believe that nothing is coincidental because everyone and everything are interconnected by consciousness. What we attribute to coincidence is really an affirmation of the "a-causal principle of connection", as Swiss Psychiatrist Carl Jung termed it, which proves, for me, the existence of a Higher Power that connects us all. I'm sure if you look back through your life, you will see that it is also defined by auspicious 'coincidences' because everyone I speak to seems to have at least one major synchronicity story, which suggests there is something way bigger going on than we can comprehend.

The following story is one such demonstration of synchronicity in my life: In August 2017, I was on holiday in Andalucía, Spain. I found details online for the one Twelve Step meeting per week on a Tuesday evening, so I made plans to attend. At the first meeting there were four people including myself, which is where I met Andy. The following Tuesday only Andy and I attended the meeting, so I asked him to share his experience, strength and hope story of recovery but to concentrate primarily on the 'Higher Power' aspect of his experience, as I was experiencing a loss of faith in my own Higher Power. I had begun to question my spiritual practices, and as a result, my prayers felt monotonous, meditation seemed pointless, and my morning readings felt meaningless. Andy narrated his extremely uplifting story, during which he mentioned being stationed in Las Palmas in Gran Canaria (a well-known seaport for the shipping industry in which he worked). I thought this was coincidental, as four years prior I had lived in Las Palmas for a few months with my Italian friends. When Andy had finished sharing, I responded with my own recovery story, and mentioned that I had also spent time in Las Palmas but I forgot to elaborate that I'd met a guy there, Martin (now a good friend), who had given me lifts in his red Seat hire car up-and-down the island to various Twelve Step meetings. Before Andy and I said goodbye to each other we exchanged details and agreed to 'add' each other on Facebook when we got home. When I arrived at the villa, I said a prayer of thanks to my Higher Power for the experience. I later discovered Andy had gone home and similarly thanked his Higher Power for the experience, which he had also found spiritually enriching. Moments later I was astonished to learn that Andy and my friend Martin were also friends on Facebook, and as I soon discovered – by messaging with them both – they had been good friends for over ten years! Andy had actually stayed with Martin on his ship, *Polar Star*, in 2014, not long after I'd returned to London from Gran Canaria.

Another such demonstration of 'spooky' synchronicity in my life is as follows: In August 2018, in the small village of Katelios on the Greek Island of Kefalonia, where I was on holiday with my fiancée Adele, I had just finished reading the *King James Bible*, which I reluctantly read for spiritual research purposes. Despite my belief in a Higher Power, having read the *Bible*, I could initially see no real value in accepting the stories to be 'the gospel' truth, as they only served to create confusion in my mind because they defy logic for three fundamental reasons. First, they are word of mouth accounts that people began to write down no less than 40 years after the death of Jesus. We all know how biased, diluted and exaggerated stories can become days or even hours after an event, and the human memory is wholly unreliable and prone to excessive imagination. Second, the valid and valuable, moral and spiritual teachings are orbited by elaborate tales of miracles that stretch the average person's believability to the limit – some of which specify raising people from the dead. Yet these types of miracles have never since been recorded by modern technology. Third, 'miracles', especially raising people from the dead, must be witnessed by multiple parties simultaneously and recorded by modern technology in a controlled, clinical environment to be scientifically proven beyond reasonable doubt. The experiments must then be repeated and observed by other parties in conjunction with the 'scientific method', as nothing less is acceptable in our secular, scientific society, and this is yet to happen. Contemplating all of this, I stood facing a lush green field peppered with white dandelions and

I implored out loud, "God, I don't believe that Jesus was raised from the dead. If I'm to believe this actually happened, I need a clear sign and I need one immediately!" I then turned to my right to walk up the hill and no more than five seconds later a man appeared from over the horizon running towards me. The man wore dark blue shorts, black running trainers and no shirt. He had olive skin, long dark-brown hair, and a big dark-brown beard – looking remarkably how Jesus is typically depicted. As the man ran past me, he greeted me with,

"Kalimera [Good Day]," and a nod and smile. Knowing inherently there is no such thing as coincidence, I knew deep within me that this was my sign from God/consciousness but doubt kicked-in and I thought, "That cannot be it! It was way too quick, too simple and too open to interpretation." So, I challenged God out loud again, "God, I don't know if that was my sign?! Can you give me another sign to cement that the running man was my sign?!"

Fifteen minutes later I was swimming in the hotel pool. As I took a short rest holding on to the side, the same man, back from his run, came walking down the path, looked me straight in the eyes and walked past me into the hotel. I had never seen him before, and I never saw him again. I consider myself an intelligent and rational person, but I cannot deny the ego-deflation and humility I felt as a result of those two 'God-instances' in quick succession. Having asked for, and been given, two signs as proof that God can perform the ultimate miracle of resurrecting the dead, my tepid belief in a Higher Power evolved to absolute faith in God/consciousness.

Make of this story what you will, but what are the chances of that happening at that particular time... twice? I was told in my early recovery that remaining sober hinged on my belief in a Higher Power. In that moment, I understood that faith is both personal and experiential, and I surrendered my logical reasoning to accept that God's miracles can destroy the wisdom of the wise and frustrate the intelligence of the most intelligent people. My recovery has benefitted hugely as a result of becoming 'God-conscious'. There isn't a moment during the day now when I can't remind myself that God is the One truth and All Is God.

Chapter 6

God

There is no empirical evidence that a God created our world. Science has never found any proof, nor has there ever been a universally convincing philosophical argument. Yuval Noah Harari wrote in 21 *Lessons for the 21st Century*, "Does God exist? That depends on which God you have in mind. The cosmic mystery or the worldly lawgiver? Sometimes when people talk about God, they talk about a grand and awesome enigma, about which we know absolutely nothing. We invoke this mysterious God to explain the deepest riddles of the cosmos. Why is there something rather than nothing? What shaped the fundamental laws of physics? What is consciousness, and where does it come from? We do not know the answers to these questions, and we give our ignorance the grand name of God. The most fundamental characteristic of this mysterious God is that we cannot say anything concrete about Him."

On one thing I think everyone can agree, if God exists It must be absolute, and this means: eternal, timeless, beyond masculine or feminine (oneness), omnipotent, omniscient, and omnipresent. Anything separate and distinct would not satisfy this criterion. If there was a God – and a universe that God created – then there would be two things, and God could not then be considered absolute. If God is absolute then nothing can be separate from God. Consciousness fits this criterion. 'Matter' is the label we give to physical phenomena, and all physical phenomena are forms of energy; consciousness manifesting at different frequencies. We make the mistake of continually looking for God outside of ourselves, presuming ourselves to be the viewer seeking God as the object of our search, but everything we perceive is tainted by our subjectivity, and anything we define as God 'out there' cannot be God because it is not absolute. It is just something that exists in relation to our perceptions. We are God. The only way to confirm this it to replace the word God with consciousness and remove the barrier of subjectivity that prevents us from our essential 'oneness' with all things. Physics is untouched by this hypothesis, as everything that is true in physics remains true. All Is One.

God, therefore, is consciousness/the ZPF/Brahman/Tao that manifests matter in an ever-progressing complexity, which is how It experiences Itself; the known universe. God is both positive and negative because the positive cannot exist or be understood without the negative. Love cannot exist or be understood without fear. Peace, joy, and compassion cannot exist or be understood without anger, misery, and hatred. I would argue that the fundamental goal of being a human is to know thyself and to transcend our negative attributes to experience love, bliss and serenity; the true nature of God. Everything in nature – plants, birds, animals, and humans – are inherently positive because they are inherently God. Just how the nature of water is wetness, the nature of plants, animals and mankind is consciousness.

Listening to theoretical physicist, mathematician, and string theorist, Professor Brian Greene, on *The Joe Rogan Experience* (podcast episode #1428), I was almost convinced by Greene's confidence that there is no God until he shared that he was terrified of deprivation tanks. I can honestly say that I am not terrified of anything. The only things I'm remotely scared of are things that can kill me, such as poisonous spiders and snakes that illicit a visceral and automatic fear response. Generally, my fear-for-no-reason has been eliminated by faith, which was born from many spiritual experiences. This is why, on careful inspection, the atheistic approach is flawed when it comes to living a fear-free life. Even British ethologist, evolutionary biologist, author, and hardcore atheist, Richard Dawkins, is scared of reputedly haunted houses according to Greene, which is paradoxical because Dawkins isn't supposed to believe in anything supernatural! In conclusion: no faith equals lots of fear.

Let's face it, the order, chaos and paradoxical contingency found in the universe presents a real puzzle. We can certainly observe a 'cosmic harmony', but capturing the elusive 'Source' of this harmony and the synchronicity of existence itself seems almost impossible. There are scientists out there, however, who reject the mainstream Big Bang theory of existence and present alternative quantum-phenomenological explanations of existence and the extension of the galaxies. Both science and the real spiritual miracles that happen in life are key to understanding these explanations. German philosopher Friedrich Nietzsche's philosophy should also be appreciated for its rejection of determinism and celebration of the freedom inherent in life's unfolding. The determinist theory states that free will is non-existent and that every decision one makes, and every thought one has, is the result of some previous situation or cause. Everything that happens, therefore, from the Big Bang – to the food a woman fills her fridge with – is determined by some preceding event. The questions then follow: how can free will exist if every choice we make is determined by a preceding event? And does free will exist or is it an illusion?

Polish American philosopher Anna-Teresa Tymieniecka's 'phenomenology of life' theory links the advancing self-individualization of life to the Logos (divine reason implicit in the cosmos that gives it form and meaning), which informs this creative development. She saw in the phenomenon of achievement – both scientific and spiritual – a sign of the self-creative activity of consciousness (known as ontopoiesis). Hence, it is imperative that scientists and spiritual scholars work together to reach an understanding of consciousness and free will. Physicist Sir Roger Penrose's Orchestrated Objective

Reduction theory (OORT) provides an indeterministic approach to consciousness, which allows for a decision-making process that, in its source of origin, can be random. He suggested that the quantum nature of thought vibrations allows free will to exist in the mind. When particles are in a quantum state, they can be in many different states at once, but when the observer views those particles, they 'choose' a state; this is quantum coherence. This is why OORT can possibly provide a space for free will to exist. If consciousness holds its roots in some sort of quantum mechanism, which I believe it does, then the quantum nature of that mechanism allows for a more random decision-making process to occur.

At this stage in the book, I'm relying on your acceptance that what religion and spirituality refer to as God/Brahman/Tao is what science refers to as the zero-point field (ZPF) of ubiquitous, quantum, matter-manifesting, energy-consciousness – and that consciousness is driving evolution indeterministically because the ability for free-will is encoded quantum mechanically in the ZPF. I feel that this is a very rational and scientific proposal of consciousness.

Many great psychologists and philosophers have proposed alternative and obscure theories of consciousness that are also worth considering. American psychologist, Timothy Leary, proposed one such theory known as The Eight-circuit Model of Consciousness, which was later expanded on by American author and mystic, Robert Anton Wilson, and astrologer, Antero Alli. The Eight-circuit Model of Consciousness suggests the brain has eight circuits operating within the human nervous system, each corresponding to its own imprint and direct experience of reality, and 24 stages of neurological evolution; three stages for each circuit that detail developmental points for each level of consciousness. The term 'circuit' is equated to a metaphor of the brain being computer hardware, and the wiring of the brain as circuitry. The first four circuits deal with life on

Earth, and survival of the species. The last four circuits are post-terrestrial, and deal with the evolution of the species, altered states of consciousness, enlightenment, mystical experiences, psychedelic states of mind, and psychic abilities. The proposal suggests that these altered states of consciousness are recently realized but not widely utilized. Leary described the first four as "larval circuits" necessary for surviving and functioning in a terrestrial human society, and proposed that the post-terrestrial circuits will be useful for future humans who are genetically hard-wired to continue to act on their urge to migrate to outer space and live extra-terrestrially. It's a very interesting theory and one that speaks to my fantasy of witnessing interstellar space travel, and hopeful contact with extraterrestrials during my lifetime.

Speaking of extraterrestrials, there is increasing interest in the Ancient Alien theory as proposed in 1968 by Swiss author, Erich von Däniken, in his seminal book, *Chariots of the Gods? Unsolved Mysteries of the Past*. Von Däniken posited a variety of hypotheses dealing with the possibility of extraterrestrial beings influencing ancient technology. Von Däniken stated that every culture in existence has ancient stories about 'gods' descending from the skies to impart wisdom, and that some ancient structures and artefacts appear to represent higher technological knowledge than is presumed to have existed at the times they were manufactured. He maintains that these artefacts were produced either by extraterrestrial visitors or by humans who learned the necessary knowledge from them. Such artefacts include the Egyptian pyramids, Stonehenge, Moai of Easter Island, and the Nazca Lines in Peru. The book also suggests that ancient artwork throughout the world could be interpreted as depicting astronauts, air and space vehicles, extraterrestrials, and complex technology. Von Däniken describes elements of artwork from unrelated cultures that he believes are similar, such as the ancient Japanese Dogū figurines (which he believed to resemble astronauts in spacesuits) and 3,000-year-old carvings in the Egyptian New Kingdom Temple that appear to depict helicopter-like machines. The book goes on to suggest that the origins of religions, including interpretations of the Old Testament of the Bible, are reactions to contact with alien races. According to von Däniken, humans considered the technology of the aliens to be supernatural and the aliens themselves to be gods. Von Däniken posits that the oral and written traditions of most religions contain references to visitors from the stars and vehicles travelling through air and space, which should be interpreted as literal descriptions that have changed during the passage of time and become more obscure. Examples include: Ezekiel's vision of the angels and the wheels, which von Däniken interprets as a description of a spacecraft; the Ark of the Covenant, which is explained as an alien communication device; and the destruction of Sodom by fire and brimstone, which is interpreted as a nuclear explosion. Von Däniken attempts to draw an analogy with the 'cargo cults' that formed during and after World War II, when once-isolated tribes in the South Pacific mistook the advanced American and Japanese soldiers for gods. He also spends around one-third of the book discussing the possibility that humans could theoretically offer primitive civilizations on interstellar worlds advanced technology by the year 2100, which would mimic the ancient extraterrestrial contact von Däniken believed to have happened on Earth. I'm not saying all this is necessarily true but, believe it or not, I've actually seen two UFOs, so my mind is open to the ancient alien theory. My first sighting was in Oldham in July 2004 with two work colleagues, who both saw a saucer-shaped object fly over an old cotton mill that we were driving alongside. Articles appeared the next day in the local newspapers – reporting many sightings of the UFO – and this information can be found on Google to this day. The second sighting was on top of Roque Nublo in Gran Canaria in 2014 with my Italian friend Massi. We both saw what looked like a star in the distance zipping across the sky – doing impossible aerobatic manoeuvres – travelling huge distances in the blink of an eye for about 30 seconds before it vanished. We were understandably dumb-founded by this exhibition of technological supremacy.

There is also mounting evidence of older and older civilizations including Tell es-Sultan in Jericho, Tell Qaramel in Syria, and the ruins of Göbekli Tepe in Turkey (which some scientists have speculated to be the first astronomical observatory or even the site of the biblical Garden of Eden). Graham Hancock discusses these ancient sites in his book, *Magicians of the Gods: The Forgotten Wisdom of Earth's Lost Civilisation*. The evidence revealed in Hancock's book shows, beyond any reasonable doubt, that an advanced civilization flourished during the Ice Age – between 12,800 and 11,600 years ago. Göbekli Tepe predated humanity's oldest known civilizations – it even seems that construction on some parts of the structures might have begun as far back as 14,000 or 15,000 years ago. Its megalithic temples were cut with laser precision from ginormous rocks millennia before the 4,500-year-old pyramids in Egypt, 5,000-year-old Stonehenge in England, or 7,000-year-old Nabta Playa in the Nubian Desert near Egypt (the oldest known astronomical site). The advanced civilization that built Göbekli Tepe was destroyed in a global cataclysm caused by a giant comet that entered the solar system from deep space thousands of years earlier and broke into multiple fragments – some of which struck the Earth on a scale unseen since the extinction of the dinosaurs. This happened on the exact date that Ancient Greek philosopher, Plato, gave for the destruction and submergence of Atlantis, and overthrows the mainstream scientific theory that humans only began building complex religions, societies and structures after the invention of agriculture. Science has had it wrong for many, many years and during that time anyone who questioned the mainstream belief, such as Hancock, were considered a pseudoscientist.

God

You might be asking yourself, what does all this have to do with God and consciousness? The answer is... everything. There were survivors of this global cataclysm – known to later cultures by such names as the 'Sages', the 'Magicians', the 'Shining Ones', and the 'Mystery Teachers of Heaven'. They travelled the whole world in their great ships doing all in their power to keep the spark of civilization burning. They settled at various locations where they built great pyramids (for reasons yet unknown) such as: Baalbek in Lebanon, Giza in Egypt, ancient Sumer in Mesopotamia (Iraq), Mexico, Peru, China, Alaska, America, and across the Pacific in Indonesia where a huge pyramid was discovered. Everywhere they went these 'Magicians of the Gods' built pyramids and brought with them the memory of a time when mankind had fallen out of harmony with the universe and paid a heavy price. Thousands of years later, I feel we have again fallen out of harmony with the universe and might be about to pay a heavy price. It's down to each individual to get in touch with their own consciousness through the practice of meditation and/or the use of psychedelic technologies in order to start turning things around. The buck stops with you and me right now.

Through the findings of modern physics, such as the theory of relativity and quantum mechanics, we now understand the duality of matter-energy, space-time, and particle-wave. We know that matter and energy are interchangeable; that space and time are relative and connected; that nature is quantum mechanical, and that quantum events have the attributes of both particles and waves. The two primary aspects or components of reality that the universe reveals to us through these discoveries are energy and consciousness. All other things that are found in the universe can be understood as manifestations of these two fundamentals. There are historical views that see these two as one. In the Eastern traditions, energy is understood to be the origin and reality of the universe; energy and consciousness are

73

viewed as different aspects of the same thing. As seen in the ancient Taoist principle that mind creates energy and energy creates mind, energy and mind are viewed as one inseparable entity. While interchanging to each other and producing limitlessly diverse phenomena, energy and mind are ultimately one. The Vedic tradition of India teaches the same principle, using different terminology: silence and dynamism are together forever. Shiva (silence, unbounded pure consciousness) and Shakti (dynamism, creativity) are always united in a cosmic embrace of wholeness that creates and sustains the world. All Is God. All Is One.

The unity of energy-consciousness is what I have realized to be the true nature of reality through my own contemplative investigations. From the time I was a young boy asking myself, "Who am I and why am I here?" I've always had a desire to know the answer. Because the desire for the answer to this question was always so compelling, I found it extremely difficult to 'fit-in' to society. When I was not able to escape from this question any longer, I went on an ayahuasca retreat in 2017, then a ten-day Vipassana silent meditation retreat in 2018 where everything became clear. In the absolute quietude and clarity of meditative solitude, I could see what I really am. The answer I had been seeking and longing for was obvious, compelling, and indisputable: the cosmic energy-consciousness is both God and my energy, and my energy is both the cosmic energy-consciousness and God. It was direct wisdom, not just knowledge. This energy-consciousness doesn't have any shape or borders and is not bound by space and time. Paradoxically, the closest word that we can find in our language to describe this enormously great, infinitely powerful, all-knowing reality that creates, sustains, and regulates the cosmos is 'Wu' (in Taoist philosophy), which translates to 'nothing'. It is not a thing nor an object. It has no qualities or characteristics. It is the un-manifest source of everything. Nothing, as the ultimate reality, is not only my personal realization of what I really am, but also the essential teaching of Taoism, to which I am greatly indebted. All this from the simple practice of meditation.

Chapter 7

Meditation

Over the last 50 years, meditation practices – generally inherited from various religious traditions – have become widely accepted as beneficial for the management of stress, and have been increasingly adopted by the mental health community (including GPs and substance misuse organisations like the one I currently work for) as a treatment modality for anxiety and depression. In light of this, you would think that psychologists and psychiatrists would be well versed in meditation and mindfulness, but the reality is they are barely touched upon when new recruits are trained in these fields – and practitioners are given very little guidance on how to respond when patients raise religious or spiritual concerns. Since meditation can be practised without reference to the religious traditions that transmitted the techniques in the first place, practitioners should feel comfortable recommending it to their patients. However, when it comes to relating to the deeper questions about life, such as the nature of suffering and the inevitability of death, practitioners generally have little idea how to proceed in my experience.

Since mental health services have no agreed-upon body of knowledge about the relationship of religion and spirituality to mental health and mental illness, it makes sense that those of us who present ourselves as having something useful to say should establish our credentials. I have been clean and sober since 5th August 2010 and I've worked in the substance misuse field as a drug and alcohol counsellor since 2015 and a Team Leader since 2017. I've experienced a ten-day Vipassana silent meditation retreat and I've experienced the power of psychedelic medicines in recovery. I have hosted the *Life in Recovery Podcast* since 2018, and this is my fifth book on the subject of spirituality. I meditate for a minimum of twenty minutes each day and have done since 2013, and I continue to study spiritual traditions, psychology, and neuroscience when I'm not working, helping others in recovery, and helping Adele raise our family.

What I've learned from meditation is that I don't need to accept anything on faith, nor accept any teachings that I can't confirm by my own personal experience. I don't accept religious teachings on face value; I try to treat them scientifically, as I do with all my experiences. Everything I write, therefore, is derived from what I've learned through study and personal experience. I'm not a scholar or teacher in these matters, nor can I pretend to be a highly accomplished meditator. Yet, all my experiences have been transformative on a personal level, giving me the capacity to cope with extreme stress at various points in my life with a degree of equanimity that would not have been possible previously. They also had a profound effect on my work as a counsellor, sponsor, and manager – in particular giving me a wider context of compassion in which to understand people's suffering. I've found that when anyone brings up religious or spiritual concerns, I'm completely at ease conversing with them in a way that is appropriate for that individual, which would have been impossible prior to becoming a meditator. For example, when I dealt with psychotic prisoners/patients in UK prisons reporting apparent supernatural experiences, I found that I could often understand what it was they were trying to describe and respond in a reassuring way that acknowledged their experiences. There is great value in being able to meet people where they are at. I ascribe whatever ability I've developed to deal with these issues effectively to a combination of my own personal spiritual experiences and my meditation practice. Equally, I ascribe my ever-progressing compassion to having developed an intellectual framework through study (especially reading such books as William James' *The Varieties of*

Religious Experience and Anthony de Mello's *The Way to Love*) for understanding the nature of spiritual experiences – and their relationship to our mental state and mental illnesses.

Many people say that they meditate but upon further inspection most of them are talking about guided meditations, which are really akin to hypnosis. Alternatively, they might chant a mantra or do yoga. These activities can be very beneficial for stress management (and sleep) but they are not meditation. Mindfulness meditation has robust stress-management benefits but that is more of a side effect. Via self-enquiry, its primary purpose is to mitigate our habitual mental patterns, which in Western psychology we have conceptualized as 'neurosis'.

There are two types of meditation practice: concentration techniques and mindfulness-awareness techniques. Concentration techniques involve focusing one's mind on a single object, such as a candle, a sound, a part of the body, or a 'mantra'. If practised sufficiently, this type of meditation can produce trance states of altered consciousness that can be intense and at times ecstatic. The intensity of these experiences is often interpreted as evidence of 'real' spirituality and can motivate people to pursue them even more enthusiastically. There are numerous traditions that utilize these types of practices. However, due to the power of the mind, there are distinct dangers in taking these techniques to extremes, and they should only be practised under the close supervision of a knowledgeable guide or 'master'. Mindfulness meditation, on the other hand (Vipassana), can be described as a simple technique for observing our own mental processes in granular detail. The typical instruction is to sit upright without a backrest to keep you awake, as meditation can be very boring. The tradition is to sit cross-legged on a cushion but there is no particular reason to adopt this 'lotus posture' or anything else uncomfortable. It's not about overcoming pain, therefore, there is no problem with sitting on a chair if needs be. Most traditions instruct people to close their eyes but some people meditate with their eyes open to prevent them from getting lost in their thoughts. Either way is fine.

The meditative technique itself is usually to turn one's attention to the breath. The problem is, after a few seconds, one's attention is lost to infinite mental chatter; the ego-mind spinning around like a washing machine. There are numerous variations and nuances in the instructions given by different traditions and different teachers on how to handle the 'washing machine' quality of our minds. The technique I learned was that as soon as we realize we are not focused on our breath, we literally say to ourselves, "Thinking," and gently return attention back to our breath. There is nothing mystical about concentrating on the breath, it's just a convenient way to help us stay in the present moment. As for answering the question of how intensely to focus on the breath, it should be "not too tight and not too loose", according to a traditional teaching of the Buddha. If we don't focus at all on the breath, no progress will be made, but if we focus too intensely, it can turn into a concentration technique, which is definitely not what is intended. If that starts to happen, it's time to lighten up, maybe look around a little and shift our position, then bring our attention back to the breath.

There is no such thing as not being able to meditate. As we move around in the world and encounter things, or as we sit in meditation and encounter things in our mind, there are three reflexive impulses that can occur. If the object or thought makes us feel good, affirmed, or safe, we want to pull it in, build it up and make it last longer. If it is threatening or uncomfortable, we want to push it away or extinguish it. And if it is neither confirming nor threatening, we ignore it. Consequently, we tend to ignore 99% of everything that crosses our awareness. In traditional Buddhist meditation texts these three impulses are known as the Three Poisons: Passion, Aggression, and Ignorance, which are the obstacles to successful meditation practice. The meaning of these three words in daily life is obvious: Passion is love, greed, obsession and addiction; Aggression is anger, cruelty and destructiveness; Ignorance is a lack of knowledge and understanding. From an evolutionary psychology perspective, these three impulses are completely natural and highly adaptive. We are attracted to what makes us feel good, we are repelled by what makes us feel bad, and we don't waste energy on the rest. In this case, however, we are not talking about our behaviour in the world. Rather, we are talking about these three impulses, as our reflexive reactions to each moment-to-moment thought. What we are being instructed to do when we sit down to meditate is very simple, very difficult, and quite unnatural; we are being asked to do none of the above. Whatever thought comes along, we are instructed to neither cultivate it, nor drive it away, nor ignore it. We are asked to simply notice it and come back to our breath.

Initially, like most people, I was terrible at meditating. I got lost in a sexual fantasy, or a business plan, or an argument with a friend or foe. I obsessed about times I had been humiliated and what I should have said, but as I persisted, little by little, I began to relax into my 'seat' (consciousness) detached from thoughts – just watching them come and go, without building them up or pushing them away, and without ignoring them. This process can be characterized as developing an attitude of equanimity toward our own thoughts. Over time, somewhat magically, I noticed that when emotional 'stuff' came up for me I was less reactive. Something that would have left me upset all day became less of a big deal. Compulsions that I would ordinarily find irresistible became less compelling. I began to feel calmer overall, clearer, and I now have more freedom to deal with what comes up in whatever way seems best – as opposed to endlessly repeating dysfunctional habitual patterns. These emotional and behavioural changes were and still are gradual and spontaneous.

I view meditation like doing chin-ups. If we start doing chin-ups but we can only do one, eventually if we have patience and practise chin-ups every day, we can do lots of chin-ups. That's why it is called meditation *practise*. Just as exercise develops your muscles, practising presence through meditation strengthens neural connections. Meditation stimulates your brain, which helps prevent atrophy and functional decline. Positive effects have been seen in long-time meditators who practise for as little as ten minutes per day. Meditation is a very simple training program, the goal of which could also be described as the art of not taking our own thoughts too seriously. This is why anyone can find meditation beneficial and the benefit is not dependent on adopting any particular philosophy, religion, or spiritual teaching.

Mindfulness meditation seems to slow the natural reduction of brain tissue that comes with aging and improves cognition and memory. Harvard researcher, Sara Lazar, found that 40-to-50-year-old meditators have key brain structures similar to those of non-meditators in their 20s; grey matter increases in the prefrontal cortex, improving focus, problem-solving and emotion regulation. The limbic system – a complex network of the brain responsible for behaviour, emotions, and survival instincts – is also altered with meditation. The hippocampus thickens, which boosts working memory and keeps you anchored in the present. Meanwhile, activity decreases in the amygdala, lessening the brain's fear response. The brain is also capable of naturally creating key chemicals such as serotonin, which helps regulate mood, social behaviour, appetite, sleep, memory, and more. Some antidepressants work by increasing the usable levels of serotonin in your brain. Mindfulness meditation has been shown to increase serotonin levels, without the side effects, making it a promising complementary therapy. Research also suggests that meditation can be effective as an additional therapy for depression, anxiety, trauma, chronic pain, cancer, heart disease, and more. Better yet, it can help prevent many issues before they start. Meditation essentially cultivates the neurobiology of optimal health and resilience.

The mindfulness-meditation model is very simple, yet very practical and effective. If one wanted to go even deeper, to understand the true nature of one's own being, there is a technique that can be added to mindfulness – known as personal enquiry. Thoughts never go away until you begin to enquire where the thoughts are coming from in the first place. One's focus can change from observing thoughts and not getting attached to them, to ending thoughts by asking the simple question, "Who am I?" Spiritual Guru, Sirshree (The Tej Gyan Foundation), wrote in *Who Am I Now*: "Self-enquiry is an ancient method of Self-realization in which you ask yourself, 'Who am I?' This method has been forgotten with the passage of time and revived by realized souls such as Guru Vashishta and Adi Shankaracharya. Today many have taken to this path after it has been strongly revived by Ramana Maharshi in the recent past... While 'Who am I?' is the question traditionally asked in Self-enquiry, 'Who am I now?' is another powerful question that further makes the practice of Self-enquiry immediate and sharp. With the addition of 'now' to the timeless question 'Who am I?' I have enhanced enquiry of the Self into enquiry in the Now. With self-enquiry, you discover your true nature. But, in your day-to-day activities, as you go through various situations and become angry, irritated, fearful, and so forth, you move away from your original essence. At such times, enquiry in the Now, i.e. 'Who am I now?' will help you shift to the present and to your true self. You may have been an angry or fearful person a moment back. But with enquiry in the Now, you will be reminded of who you actually are and be led back to your true nature."

With the earnest practice of questioning, such as, "What is the state of my mind right now?", "Who am I?" and "Who am I now?" your mind surrenders and you learn to live in the present moment and get established in the state of being. In that state, you come to know that your body is merely your mirror. It reflects yourself to you and you are able to experience yourself thanks to your body. If the feeling of anger arises in your body, you would say, "I am angry." But if you ask, "Who is angry?" a profound insight will emerge that it's never you that has felt angry. You were always free, are free, and will always remain free. Anger is just a passing emotion in the body that gives rise to thoughts created by your ego-mind.

Sirshree also wrote, "When you see yourself in the right way, you can achieve liberation from the cosmic illusion. Some call this state Self-realization, while others call is moksha, nirvana, samadhi, kaivalya, mukti, or enlightenment. All these terms are given to the same state, which is the state of being. This state too is called by different names such as: I-am-ness, sense of presence, self-witness, consciousness, etc... The ultimate purpose of every human being is to attain self-realization and express the qualities of the Self (which is also known as God, Consciousness, Lord, Creator, Ishwar, Allah, Witness, Experience of being, etc.). The Self is formless and limitless, it's omnipresent and omnipotent. But the characteristic of the Self is that it connects with something and starts believing itself to be that object or being. When the Self connects with a body, it considers itself to be that limited body... The mind is also a creation of the Self."

This process of self-enquiry can lead one to the ultimate truth of enlightenment. Although I've had glimpses of enlightenment via meditation and psychedelics, which I believe show you what is on offer should you choose to put in the work and attain enlightenment through meditation, these are just highlights in a life of continuous and consistent action. Despite the common perception of enlightenment to be bliss or profound insight, it is actually a pure-consciousness event that is literally nothing. Enlightenment is entirely empty because consciousness is entirely empty. This has been my experience: following bliss comes emptiness. Yet enlightenment is profoundly satisfying and transformative despite the fact that the mind of the enlightened person remains unchanged. According to the mystics, once enlightened, you will remain neurotic and still want to have sex and buy material possessions – it doesn't shift all that stuff – but there is deep wisdom that arises that wasn't there before. Conversely, enlightenment can also lead to narcissism, as it can be the greatest power-trip ever – an aphrodisiac even. When some people have a profound mystical revelation, they think they are God, and that can have an awesome effect on people around them, who become enamoured with their 'Godliness'.

Enlightened 'Masters' say that, just as electrons can be described as both waves and particles, ultimate reality is both timeless and aimless yet has some directionality and purpose. There is a void, therefore, at the heart of reality. Not God, or ecstasy, or a theory of everything, but nothing. So why, I hear you ask, is enlightenment the most common goal of the spiritual seeker?

Chapter 8

Enlightenment

In Buddhism, the Seven Factors of Awakening that lead to enlightenment are: the cultivation of awareness of reality and the investigation of the nature of reality through mindfulness and meditation; determination and effort (daily meditation practice); experiencing joy (or rapture) as a result of mindfulness and meditation; relaxation (or tranquillity) of both body and mind as a result of mindfulness and meditation; a one-pointed state of mind or clear awareness (Samadhi) during meditation; equanimity in one's daily life as a result of mindfulness and meditation; and accepting reality *as it is* without craving or aversion. Once one has achieved this state of being, one is said to be enlightened.

Is enlightenment a mere fantasy? I don't think so. What I do know for certain is you need a burning desire to attain enlightenment, so you will continue the journey of change diligently and carry out your spiritual practices with perseverance. It's very easy, once you have experienced the Self/consciousness (and glimpsed enlightenment) to be satisfied with your achievement, but if your body has the wrong tendencies, habits, patterns, or addictions (like mine does), it will not allow you to go back into the Self/consciousness again. In fact, these obstacles go on reducing your efforts as your ego-mind gives more and more excuses. The egotistical mind-frame, "I am knowledgeable", also gets added to the list of obstacles, and you can get entangled once again in the web of ego-mind-games. I know this because I was lost there for some time until I realized, in July 2020, that I would have to drop my last remaining habitual and compulsive behaviour patterns if I was to continue to change and grow toward enlightenment. I know that once my habitual and compulsive behaviour patterns are overcome, I will constantly reside in the Self/consciousness and become like a child – like my daughter Alice – who is always in the Self/consciousness and always in the present moment: enlightened.

Enlightenment cannot be achieved, however, before one becomes wise, which means putting all your knowledge and understanding into practice all of the time. Not pressing the 'fuck-it' button and acting out with substances or behaviours when something doesn't go your way is the first step. The second step is learning to pause (as Self/consciousness/God is in the pause) and not react to situations (don't get in the ring as my sponsor always says); essentially meditating for a few moments and allowing emotions to pass or settle. This is doable once you have the understanding that emotions and thoughts emerge continuously, often randomly, from consciousness – and we have no control over them. The third step in becoming enlightened is to act appropriately in all circumstances, all of the time.

Knowledge and understanding of self (and therefore others) is essential to our evolution. Knowledge and understanding is a precursor to wisdom, without which Homo Sapiens cannot, and will not, propagate. As the RZA puts it in *The Tao of Wu*, "Wisdom is the Light. Wisdom is what shows those in the darkness the Light, what reveals the path or the Way. It's what we need to live. The sutras of the Buddha teach that without wisdom there is no gain. In the Bible's Book of Proverbs, King Solomon chooses wisdom over all the other gifts that God offers him. In Islam's Divine Mathematics, we learn that wisdom is proof of knowledge, reflection of knowledge, knowledge in action. In my life, all these understandings of wisdom have proven true."

Once you are truly wise, you can always see the truth and your actions will never cause harm to yourself or others. It is the end of suffering for you and those around you. The Buddha said, "Enlightenment is the end of suffering." He did not mean physical suffering, as there is no way to avoid physical pain. Even Jesus cried out in pain during his crucifixion. The suffering that Buddha referred to was the suffering caused by the sense of 'doership': pride and arrogance for one's good actions, guilt and shame for one's bad actions, and hatred towards the other for their actions. Accepting that everything in existence is consciousness experiencing Itself (God's will) removes the burden of pride, arrogance, guilt and shame, and what remains is peace of mind.

Wisdom is fluid. It brings flexibility and adaption. It frees you from slavery to your past and your obsessions. Twelve Step recovery was my path to wisdom. Through doing 'Step-work' I gained knowledge of myself and began putting that knowledge into practice. Over the years, the gathering of knowledge has evolved my understanding of what it is to be a human, and enabled me to love myself and others unconditionally – because there is no 'other'. Thanks to wisdom, I now experience anger the same way I experience joy; I feel it then I let it pass. Like all thunderstorms, it always passes. Resentments only affect me for a couple of hours at most before they pass. There is only the Self/consciousness experiencing Itself. This is the ultimate wisdom that arises from an ever-progressing awareness of the material universe through conscious experiences.

In a 2017 essay titled, "Minding matter", professor of astrophysics at the University of Rochester, Adam Frank, eloquently expresses both the mystery of consciousness and the reluctance of scientists to propose theories that oppose viewing consciousness as a result of brain processing: "After more than a century of profound explorations into the subatomic world, our best theory for how matter behaves still tells us very little about what matter is. Materialists appeal to physics to explain the mind, but in modern physics the particles that make up a brain remain, in many ways, as mysterious as consciousness itself... Rather than trying to sweep away the mystery of mind by attributing it to the mechanisms of matter, we must grapple with the intertwined nature of the two... Consciousness might, for example, be an example of the emergence of a new entity in the universe not contained in the laws of particles. There is also the more radical possibility that some rudimentary form of consciousness must be added to the list of things, such as mass or electric charge, that the world is built of."

I wholeheartedly concur with Professor Frank. The 'first cause' of everything is consciousness in my opinion. There is never anything but consciousness in varying forms – happening right now in the present moment. If one could view the universe objectively from a 'God's-eye-view', in the same way a human might view the inside of an anthill, one would see everything happening at once, as time is not linear. Time is simply a human construct intended to structure every now moment. American author, Annaka Harris, wrote in *Conscious: A Brief Guide to the Fundamental Mystery of the Mind*, "Many neuroscientists have considered the possibility that the feeling we have of being in the present moment, with time continuously moving in one direction, is an illusion. In his book, *Your Brain Is a Time Machine*, Dean Buonomano, a UCLA neuroscientist, explains that whether the flow of time is an illusion or a true insight into the nature of reality depends in part on which of these two opposing views in physics turns out to be correct:

1. Presentism: Time is in fact flowing and only the present moment is 'real'; or
2. Eternalism: We live in a 'block universe,' where time is more like space – just because you are in one location (or moment) doesn't mean the others don't exist simultaneously."

I personally subscribe to this notion of a 'block universe', meaning the past, present, and future all exist at the same time. There really are no 'events', there is just one single event or 'datum'. Causality is really an illusion, which is why theoretical physicist Albert Einstein said, "The distinction between the past, present and future is only a stubbornly persistent illusion," and quantum theorist Christopher Fuchs said, "Maybe that's what quantum mechanics has been trying to tell us all along – that a single objective reality is an illusion."

The past is merely a recollection of now moments that we call a memory in the present (and our memories are not very reliable, so it is a loose recollection at best); the past is just theory. We can say whatever we like about the past because it doesn't actually exist. In other words, you choose your past based on your five sense perceptions. Conversely, the future is simply an imagining of what may or may not happen – and it rarely happens how we imagine it will anyway. Therefore, all that ever exists is now... and now... and now... ad infinitum. When I say, "All Is One", and "All Is Now", this really is the truth, which is why enlightened Sufi mystic Rumi said, "Past and future veil God from our sight; burn them both up with fire."

Quantum physics has thrown 'reality' into question by proving that, without an observer, 'now' does not exist, therefore the universe does not actually 'exist' as we perceive it. Reality *is* an illusion. The evolution of the universe and everything in it, including life itself, is possible only relative to an observer-participant. Without an observer, we have a dead universe, which does not evolve in time, implying the vital role of the participant in the self-observing universe. Studies in quantum mechanics have thrown up astonishing results, such as particles that can be in two states at once, and the behaviour of the particles seems to change depending on whether they are being measured or not (challenging the idea of an objective reality). In fact, even an intention to measure them seems to change their behaviour. Advocates of quantum theories of consciousness, like British physicist Roger Penrose, believe this implies consciousness is somehow linked with the quantum world, and that quantum processes in the brain could help to explain consciousness.

To be enlightened is to know that the physical world emerges from an enmeshed unity of space-time, primordial consciousness, and a primal energy that is indivisible from both space and consciousness. This ultimate reality exists in the 'fourth time' (the ZPF), beyond distinctions of past, present, and future. Primordial consciousness is the formless realm of numbers that gives rise to the realm of geometric forms, which gives rise to the realm of physical space, time, energy and matter that gives rise to the realm of corporeal sentient beings, which gives rise to the realm of primordial space, consciousness and life force, completing the circle back in the formless realm. When the universe did not exist, only consciousness existed. Consciousness created the universe only for experiencing Itself. This world is a mirror for consciousness to witness Itself and the whole universe is a medium for consciousness to experience and express Itself. Consciousness is no-thing, with the potential of everything; it cannot be fully expressed in words. It can only be experienced.

Paradoxically, consciousness is not only within us but the body exists within consciousness. All of this existence is happening within consciousness; like a fish in the ocean that does not know that it lives in water. Water exists not only within the fish but also all around it. The fish doesn't realize it is in water and never goes in search of water. From the standpoint of consciousness, there is no inside or outside. Consciousness is your very essence, It is limitless and present everywhere.

I think the most obvious yet important point I can make is that I am conscious, you are conscious, and you know every other human being is conscious – and probably every other living organism. It makes sense, therefore, to conclude that we should treat all humans and all living organisms with respect and love because at the base level we are all the same: All Is One. The *Upanishads* preach a message of unity and are opposed to any form of discrimination. To adapt the words of the *Isha Upanishad*, "Who sees all beings in their own self and their own self in all beings, loses all hatred and fear."

The alternative to this view is the predominant scientific view that consciousness is nothing more than emergent phenomena in a complex brain – in a material world void of spirit. Is that what is really happening? Is everything we feel and think limited to our brain's reactions to the world? I don't think so. Is it too much to believe that the brain of all organisms has evolved (on a progressively more complex scale) to 'tune-in' consciousness, like a radio tunes in a signal, to enable organisms to have what we call conscious experiences? In physics, revelations from the subatomic world have forced scientists to acknowledge the inadequacy of precise measurement and prediction, and accommodate a new worldview where mind and matter, fact and value, cannot be divorced. This relational worldview and the centrality of perspective within it stands in contention with the objectivity one typically associates with science. Essentially, the more complex a brain structure, the more conscious the organism; but the brain does not create consciousness, it 'tunes in' to consciousness like a TV set tunes in a picture, offering a plethora of experiences dependant on the brain structure of the organism. Various lifeforms have evolved over millions of years to be able to experience emotions, thoughts and behaviours that arise, but every organism that does not have a brain is technically conscious – in that they are imbued with consciousness yet have no way of experiencing their own consciousness.

At the time of writing, scientists from the University of Michigan proved that the brain's default mode network (DMN) and the dorsal attention network (DAT) are anti-correlated, meaning that when one is active, the other is suppressed. The team also found that neither network was highly active in people who were unconscious. These findings suggest that the interplay of the DMN and the DAT support consciousness by allowing us to interact with our surroundings then to quickly internalise those interactions, essentially turning our experiences into thoughts and memories. The relationship between the two has been studied before, but the Michigan team's research yielded the first definitive proof that the DMN and DAT are, in fact, anti-correlated. If you think about it, it makes sense. It's hard to be fully engaged with your surroundings and be deep in thought about yourself at the same time. People meditate to try to get 'out of their heads' and focus on the present moment, that is, to quiet the DMN and activate the DAT. Psychedelics like psilocybin and ayahuasca have the same effect: the default mode network is quieted, often resulting in intense feelings of connection to the natural world, other people, and one's surroundings. Since the DMN is where our egos live and where negative thought loops about ourselves take place, the use of psychedelics to quiet this brain region is increasingly being studied as a treatment for depression, PTSD, addiction, and other neurological disorders. We try pretty hard these days to get out of our own heads – and it's not easy. This study showed that not only can we not be in our own heads and out of them at the same time, but this mutually exclusive relationship between the DMN and the DAT – and the consistent switching between them – is what enables us to interact with our environment then internalise and process our experiences. In other words, to be conscious.

The researchers also saw that the brain quickly transitions from one network to another in regular patterns, and the conscious brain cycles through a structured pattern of states over time, including frequent transitions to the default mode and dorsal attention networks. However, in patients who were unconscious – whether they'd been sedated or had suffered from brain disorders – transitions to the DMN and DAT were much less frequent. Although the experiences of unresponsive patients would have differed depending on how they became unconscious – their brain networks would have been impacted and reorganised in different ways – they all shared the same isolation of the DMN and DAT networks. In people who are conscious, turning off the DMN results in an inability to deeply self-reflect. Turning off the DAT, on the other hand, results in an inability to be aware of and respond to one's surroundings. The researchers already knew that we are in a conscious state whether we are daydreaming and caught up in memories or out of our head and engaged with the world around us. These findings were further proof that, one, we don't use the brain networks required for self-reflection and external engagement at the same time, and two, we don't use much of either when we are unconscious.

So, what does this prove? Without the involvement of one's subjective self, what we humans call emotions cannot be experienced. Other animals might have some kinds of emotional experiences in significant situations in their lives, but without 'autonoesis' (the capacity to be aware of one's own existence as an entity in time) they cannot have the kinds of experiences we do. Pondering such issues, the philosopher Todd May recently asked, "Would human extinction be a tragedy?" He concluded that the planet might well be better off without us, but that such an outcome would indeed be a tragedy, as we have achieved remarkable things as a species. Autonoesis, arguably, has made these things possible. But it also has a dark side. With self-consciousness comes selfishness, and narcissism, enabling our most troubling and base dispositions toward others: distrust, fear, hate, greed and obsession. I would argue that human selfishness is the root of all evil in the world, yet, only self-conscious human minds can come to the realization, as May's mind did, that we have an obligation to confront our selfish nature for the overall good of humankind. To act on this, however, will require a global altruistic effort. If we can succeed in joining together to rise above self-indulgent desires we might avert some of the potential disasters that might await us, and preserve some kind of future for our descendants.

Altruism is difficult to account for because most scientists believe that humans are just genetic machines, only concerned with the survival and propagation of our genes. From a 'spiritual' perspective, however, altruism is easy to explain, as it is related to empathy. Human-shared fundamental consciousness means that it is possible for us to sense the suffering of others and to respond with altruistic acts. Since we share fundamental consciousness with other species too, it is possible for us to feel empathy for – and behave altruistically toward – them as well. Following an 'awakening experience', like I had in 2009, human awareness intensifies and expands, and we experience a sense of 'oneness' with other human beings, nature, and the universe as a whole. I believe awakening experiences are encounters with fundamental consciousness, in which we sense its presence in everything around us, including our own selves. We experience a sense of oneness because oneness is the fundamental reality of everything.

My own sense of the correct resolution to the mystery of consciousness, whether or not we can ever achieve a true understanding, is split between science and spirituality. Despite the progress made in studying observable conscious processes in the brain, there are still no satisfactory answers to the two core questions: what is the origin of the subjective, phenomenal aspect of consciousness? and what are the causal mechanisms underlying the generation of individual phenomenal states? A fresh outlook is definitely needed. One that is grounded in objective findings and rigorous scientifically based analysis but, at the same time, open to philosophically and spiritually subjective metaphysics, which has always argued for the base assumption of a cosmic level of consciousness serving as the ultimate bedrock of experiential reality. My answer to this is a combination of the stochastic electrodynamics (SED) framework, which theorises that all conceivable shades of phenomenal awareness are woven into the frequency spectrum of a universal background field (named zero-point field or ZPF), and panpsychism, which relies on the central idea that the universe is imbued with a ubiquitous field of consciousness. This zero-point field is a dual-aspect foundation of the cosmos, the extrinsic appearance of which is physical in nature and the intrinsic manifestation of which is phenomenological in nature. A complex brain acquires phenomenal properties (thoughts and feelings) by tapping into the universal pool of phenomenal nuances predetermined by the ZPF, or more specifically, by entering into a temporary liaison with the ZPF and extracting a subset of phenomenal tones from the phenomenal colour palette inherent in the basic structure of the ZPF. This combined theory asserts that the universe is imbued with an inherently sentient medium that is seamlessly embedded in the edifice of modern quantum physics. Consciousness is, therefore, the ultimate intrinsic force underlying the dynamic transformations described by quantum physics, thus laying the foundations for a scientifically informed idealist worldview.

In conclusion, here are ten alternative hypotheses about the origins of consciousness for you to ponder:

1. Consciousness has always existed, because God is conscious and eternal.
2. Consciousness began when the universe formed, around 13.7 billion years ago (panpsychism).
3. Consciousness began with single-celled life, around 3.7 billion years ago.
4. Consciousness began with multicellular plants, around

850 million years ago.

5. Consciousness began when animals such as jellyfish got thousands of neurons, around 580 million years ago.

6. Consciousness began when insects and fish developed larger brains with about a million neurons (honeybees) or 10 million neurons (zebrafish) around 560 million years ago.

7. Consciousness began when animals such as birds and mammals developed much larger brains with hundreds of millions of neurons, around 200 million years ago.

8. Consciousness began with humans, Homo Sapiens, around 200,000 years ago.

9. Consciousness began when human culture became advanced, around 3,000 years ago.

10. Consciousness does not exist, as it is just a scientific mistake (Behaviourism) or a "user illusion" (Daniel Dennett).

I think it's fair to say that hypothesis 7 is currently the most popular but I feel that I have presented enough evidence, both scientific and spiritual, to state that a combination of hypotheses 1 & 2 are more sufficient in explaining the universe as we know it. I propose that: consciousness has always existed because it is eternal, and the material universe formed around 13.7 billion years ago when consciousness brought it into existence in the panpsychic sense.

Perhaps the term 'panpsychism' will continue to pose obstacles to progress in consciousness studies. We might need a new label for the work in which scientists and philosophers theorise about the possibility that consciousness is fundamental, such as 'The Organizing Principle Theory'. At the very least, it seems clear that the current incomplete picture gives us good reason to keep thinking creatively about consciousness from both a scientific and spiritual perspective. And of one thing I am convinced, the universe is not one unbending unit of fact – it is always subject to change, as change is the only true constant. There is an ever-progressing complexity of diversity in it, and I don't need to be enlightened to know this. I see it all around me all the time because, thankfully, I am awake and aware, and no longer ignorant to it. I couldn't sum up my overall sentiment any better than French philosopher, Henri Bergson, who wrote:

Fortunately, some people are born with spiritual immune systems that sooner or later give rejection to the illusory worldview grafted upon them from birth through social conditioning. They begin sensing that something is amiss, and start looking for answers. Inner knowledge and anomalous outer experiences show them a side of reality others are oblivious to, and so begins their journey of awakening. Each step of the journey is made by following the heart instead of following the crowd and by choosing knowledge over the veils of ignorance.

References

1. The Role of the Brain in Conscious Processes: A New Way of Looking at the Neural Correlates of Consciousness
Link: https://www.ncbi.nlm.nih.gov/pmc/articles/PMC6085561
Author: Joachim Keppler
Publication: Frontiers in Psychology

2. Metaphysics of Consciousness
Link: https://link.springer.com/chapter/10.1007/978-981-13-7228-5_7
Author: Ramesh Chandra Pradhan
Publication: Mind, Meaning and World

3. Top Mysteries of the Mind: Insights From the Default Space Model of Consciousness
Link: https://www.frontiersin.org/articles/10.3389/fnhum.2018.00162/full
Authors: Ravinder Jerath and Connor Beveridge
Publication: Frontiers in Human Neuroscience

4. Consciousness: Metaphysical Speculations and Supposed Distinctions
Link: https://link.springer.com/chapter/10.1007/978-3-319-9363 5-2_7
Author: Grant Gillett
Publication: From Aristotle to Cognitive Neuroscience

5. Seeking the essence of consciousness in the human brain: spirituality or science?
Link: https://riverheadlocal.com/2019/03/24/seeking-the-essence-of-consciousness-in-the-human-brain-spirituality-or-science/
Author: Greg Blass
Publisher: riverheadlocal.com

6. Does Consciousness Come in Degrees?

Link: https://iai.tv/articles/does-consciousness-come-in-degrees-auid-1226
Author: William Lycan
Publisher: iai.tv

7. How Does Consciousness Work?
   Link: https://lithub.com/how-does-consciousness-work/
   Author: Michael S. Gazzaniga
   Publisher: lithub.com

8. Consciousness, Value, and Human Nature
   Link: https://link.springer.com/chapter/10.1007/978-3-319-93635-2_5
   Author: Grant Gillett
   Publication: From Aristotle to Cognitive Neuroscience

9. Consciousness Isn't Self-Centered
   Link: http://m.nautil.us/issue/82/panpsychism/consciousness-isnt-self_centered
   Author: Annaka Harris
   Publisher: m.nautil.us

10. Microtubules and Human Consciousness
    Link: https://excal.on.ca/microtubules-and-human-consciousness/
    Author: excal
    Publisher: excal.on.ca

11. IT IS EASIER TO BE "PRESENT" IF YOU CAN MASTER THIS ONE COGNITIVE TECHNIQUE
    Link: https://www.inverse.com/mind-body/scientists-explain-the-cognitive-techniques-to-use-if-you-want-to-feel-present
    Author: Sarah Sloat
    Publisher: inverse.com

12. The Evolution of Consciousness Enables Conscious Evolution
    Link: https://evolution-institute.org/the-evolution-of-consciousness-enables-conscious-evolution/

Author: Steven C. Hayes
Publisher: evolution-institute.org

13. According to the Upanishads our true nature, Spirit, is ... (a field of pure consciousness)
Link: http://adishakti.org/_/according_to_the_upanishads_our_true_nature_spirit_is_a_field_of_pure_consciousness.htm
Author: adishakti
Publisher: adishakti.org

14. Can babies see auras?
Link: https://psychologicallyastrology.com/2018/09/28/of-babies-and-auras/
Author: astrologerbydefault
Publisher: psychologicallyastrology.com

15. 365 Tao: Daily Meditations
Author: Deng Ming-Dao
Publisher: Bravo Ltd

16. 21 Lessons for the 21st Century
Author: Yuval Noah Harari
Publisher: Vintage

17. Civilized to Death
Author: Christopher Ryan
Publisher: Simon & Schuster

18. Conscious
Author: Annaka Harris
Publisher: Harper

19. The Tao of Wu
Author: The RZA
Publisher: Riverhead

20. Largest ever DMT survey travels to the fringes of psychedelic science
Link: https://newatlas.com/science/dmt-survey-psychedelic-atheism-johns-hopkins-alan-davis/
Author: Rich Harid

Publication: New Atlas

21. Drugs and the spiritual: Bill W. takes LSD.
    Link: http://www.williamwhitepapers.com/pr/1989%20Bill
    %20W%20takes%20LSD.pdf
    Author: William whitepapers
    Publisher: williamwhitepapers.com

22. Has the Playing Field Been Leveled?
    Link: https://www.psychiatrictimes.com/view/has-playing-
    field-been-leveled
    Author: Jane B. Sofair, MD
    Publisher: psychiatrictimes.com

23. Episode 5: Mind, consciousness and freewill: Are we more
    than matter?
    Link: https://www.thebigconversation.show/daniel-denne
    tt-and-keith-ward
    Author: Unbelievable?
    Publisher: thebigconversation.show

24. Prof. B. Alan Wallace talks on "Mind, Emptiness and
    Quantum Physics." #Day 2
    Link: https://www.youtube.com/watch?time_continue=16
    &v=UpxBilZeCUQ&feature=emb_logo
    Author: Library of Tibetan Works and Archives
    Publisher: YouTube

25. Three Problems With the Big Bang
    Link: https://www.realclearscience.com/blog/2016/05/three
    _problems_with_the_big_bang.html
    Author: Ross Pomeroy
    Publisher: realclearscience.com

26. Who Am I Now
    Author: Sirshree
    Publisher: WOW Publishings

27. Minding matter
    Link: https://aeon.co/essays/materialism-alone-cannot-expl
    ain-the-riddle-of-consciousness

Author: Adam Frank

Publisher: Aeon

28. Consciousness Cannot Have Evolved

Link: https://iai.tv/articles/consciousness-cannot-have-evolved-auid-1302

Author: Bernardo Kastrup

Publisher: iai.tv

29. How to Mitigate the Hard Problem by Adopting the Dual Theory of Phenomenal Consciousness

Link: https://www.frontiersin.org/articles/10.3389/fpsyg.2019.02837/full

Authors: M. Polák and T. Marvan

Publication: Frontiers in Psychology

30. Consciousness Isn't Self-Centered

Link: https://getpocket.com/explore/item/consciousness-isn-t-self-centered

Author: Annaka Harris

Publisher: POCKET WORTHY

31. KASTRUP: NO, CONSCIOUSNESS CANNOT BE JUST A BYPRODUCT

Link: https://mindmatters.ai/2020/02/kastrup-no-consciousness-cannot-be-just-a-byproduct/

Author: NEWS

Publisher: Mind Matters

32. PHILOSOPHER: CONSCIOUSNESS IS NOT A PROBLEM. DUALISM IS!

Link: https://mindmatters.ai/2020/05/philosopher-consciousness-is-not-a-problem-dualism-is/

Author: NEWS

Publisher: Mind Matters

33. Are dogs conscious? How about computers? Brain scientist Christof Koch takes on deep questions

Link: https://www.geekwire.com/2019/dogs-conscious-computers-brain-scientist-christof-koch-takes-deep-questions/

Author: Alan Boyle
Publisher: GeekWire

34. Energy and Consciousness: Something From Nothing
Link: https://brainworldmagazine.com/energy-and-conscio usness-something-from-nothing/
Author: Ilchi Lee
Publication: Brain World

35. Scientists just proved these two brain networks are key to consciousness
Link: https://www.themandarin.com.au/128731-scientists-just-proved-these-two-brain-networks-are-key-to-consciousness/
Author: Vanessa Bates Ramirez
Publisher: The Mandarin

36. Consciousness is real
Link: https://aeon.co/essays/consciousness-is-neither-a-spo oky-mystery-nor-an-illusory-belief
Author: Massimo Pigliucci
Publisher: Aeon

37. There Are Two Hard Problems of Consciousness, Not One
Link: https://www.psychologytoday.com/us/blog/theor y-knowledge/201910/there-are-two-hard-problems-consciousness-not-one
Author: Gregg Henriques, PhD
Publication: Psychology Today

38. Does Reality Shift As Our Consciousness Evolves? Ask Deepak Chopra!
Link: https://www.youtube.com/watch?v=zHTMDqjNY08
Author: Deepak Chopra
Publisher: YouTube

39. Anatomy 101: The Brain Science Behind Meditation
Link: https://www.yogajournal.com/.amp/meditation/scie nce-behind-being-present
Author: Ann Swanson

Publication: Yoga Journal
40. What Erwin Schrödinger Said About the Upanishads
    Link: https://science.thewire.in/the-sciences/erwin-schro-dinger-quantum-mechanics-philosophy-of-physics-upani-shads/
    Author: Viraj Kulkarni
    Publisher: Science: The Wire
41. Cosmopsychism and Consciousness Research: A Fresh View on the Causal Mechanisms Underlying Phenomenal States
    Link: https://www.frontiersin.org/articles/10.3389/fpsyg.2020.00371/full
    Author: William Seager
    Publication: Frontiers in Psychology

Author Biography

Ren Koi is a British author and host of the *Life in Recovery Podcast*. An ex-DJ and music producer (John Loxley aka Lox), he spent over ten years drinking alcoholically and using drugs before joining Twelve Step fellowships in 2010 following an emotional breakdown.

John had a spiritual awakening and a psychic change as a result of working the Twelve Step Program. Having a strong affiliation with Japanese culture, he adopted the pseudonym Ren Koi (Ren meaning both 'Lotus' and 'love' in Japanese, and Koi being the symbol of courage in Buddhism while also being associated with perseverance in adversity and strength of purpose in Japan) and began writing books.

In his first self-published book, *Addiction Prevention: Twelve Steps to Spiritual Awakening* (2016), Ren proposed that the implementation of addiction prevention classes in schools could prevent the onset of mental health issues that people attempt to medicate with addictive substances and damaging behaviours, which lead to addictions.

Working in the substance misuse field since 2015, Ren has tried to carry the message of recovery to still-suffering alcoholics and addicts. He works in partnership with people both personally and professionally to help them change their lives for the better, and achieve positive and life-affirming goals.

Previous Books

Addiction Prevention: Twelve Steps to Spiritual Awakening
ISBN: 9781543293395

Few people are fully aware that we are at the beginning of a
mental health crisis and addiction epidemic across the world.
It is Ren Koi's belief that we might halt the addiction epidemic
by treating emotional disorder with the Twelve Step Program,
and thus prevent the onset of mental illness that people
attempt to medicate with addictive substances and damaging
behaviours.
Part autobiography, part philosophy and part educational tool,
the book advocates the implementation of Twelve Step classes
in schools. If you hope for a better future for our children and
you are interested in self-improvement, this book is for you.

Anonymous God: Coincidence, Serendipity, Synchronicity, Spiritual Signposts and Psychedelics
ISBN: 9781985091214

Is there such a thing as coincidence, or is every coincidence
really a God-instance; synchronistic events orchestrated by a
Higher Power? Ren Koi explores these phenomena in relation
to spiritual awakening and to finding meaning and purpose in
life. The book also discusses the potential role of psychedelic
drugs in awakening mankind to the interconnectedness of our
lives, which cannot be explained by scientific reasoning.

Together: An Ayahuasca Experience
ISBN: 9781797717180
A seemingly fictional non-fiction love story with a psychedelic twist.

The Spiritual Malady: How to Attain Peace of Mind and Lasting Happiness
ISBN: 9798612262251
We are all suffering with The Spiritual Malady to a greater or lesser degree. We all feel a sense of dis-ease bordering on despair at times.
We all want to be happy but to enjoy lasting happiness, first we must attain peace of mind.
Ren Koi explains exactly how you can attain peace of mind and enjoy lasting happiness via ten simple lessons.

From the Author

Thank you for purchasing *All Is One*. My sincere hope is that you derived as much from reading this book as I have from creating it. If you have a few moments, please feel free to add your review of the book to your favourite online site for feedback. Also, if you would like to connect with me, please visit my website for my contact details, social media links and my *Life in Recovery Podcast*: https://www.lifeinrecovery.co.uk

Peace and love,

Ren

BOOKS

SPIRITUALITY

O is a symbol of the world, of oneness and unity; this eye represents knowledge and insight. We publish titles on general spirituality and living a spiritual life. We aim to inform and help you on your own journey in this life.
If you have enjoyed this book, why not tell other readers by posting a review on your preferred book site?

Recent bestsellers from O-Books are:

Heart of Tantric Sex
Diana Richardson
Revealing Eastern secrets of deep love and intimacy to Western couples.
Paperback: 978-1-90381-637-0 ebook: 978-1-84694-637-0

Crystal Prescriptions
The A-Z guide to over 1,200 symptoms and their healing crystals
Judy Hall
The first in the popular series of six books, this handy little guide is packed as tight as a pill-bottle with crystal remedies for ailments.
Paperback: 978-1-90504-740-6 ebook: 978-1-84694-629-5

Your Simple Path
Find Happiness in every step
Ian Tucker
A guide to helping us reconnect with what is really important in
our lives.
Paperback: 978-1-78279-349-6 ebook: 978-1-78279-348-9

365 Days of Wisdom
Daily Messages To Inspire You Through The Year
Dadi Janki
Daily messages which cool the mind, warm the heart and guide
you along your journey.
Paperback: 978-1-84694-863-3 ebook: 978-1-84694-864-0

Body of Wisdom
Women's Spiritual Power and How it Serves
Hilary Hart
Bringing together the dreams and experiences of women across
the world with today's most visionary spiritual teachers.
Paperback: 978-1-78099-696-7 ebook: 978-1-78099-695-0

Dying to Be Free
From Enforced Secrecy to Near Death to True Transformation
Hannah Robinson
After an unexpected accident and near-death experience, Hannah
Robinson found herself radically transforming her life, while a
remarkable new insight altered her relationship with her father, a
practising Catholic priest.
Paperback: 978-1-78535-254-6 ebook: 978-1-78535-255-3

The Ecology of the Soul

A Manual of Peace, Power and Personal Growth for Real People
in the Real World
Aidan Walker
Balance your own inner Ecology of the Soul to regain your
natural state of peace, power and wellbeing.
Paperback: 978-1-78279-850-7 ebook: 978-1-78279-849-1

Not I, Not other than I

The Life and Teachings of Russel Williams
Steve Taylor, Russel Williams
The miraculous life and inspiring teachings of one of the World's
greatest living Sages.
Paperback: 978-1-78279-729-6 ebook: 978-1-78279-728-9

On the Other Side of Love

A woman's unconventional journey towards wisdom
Muriel Maufroy
When life has lost all meaning, what do you do?
Paperback: 978-1-78535-281-2 ebook: 978-1-78535-282-9

Practicing A Course In Miracles

A translation of the Workbook in plain language, with
mentor's notes
Elizabeth A. Cronkhite
The practical second and third volumes of The Plain-Language
A Course In Miracles.
Paperback: 978-1-84694-403-1 ebook: 978-1-78099-072-9

Quantum Bliss

The Quantum Mechanics of Happiness, Abundance, and Health

George S. Mentz

Quantum Bliss is the breakthrough summary of success and spirituality secrets that customers have been waiting for.

Paperback: 978-1-78535-203-4 ebook: 978-1-78535-204-1

The Upside Down Mountain

Mags MacKean

A must-read for anyone weary of chasing success and happiness – one woman's inspirational journey swapping the uphill slog for the downhill slope.

Paperback: 978-1-78535-171-6 ebook: 978-1-78535-172-3

Your Personal Tuning Fork

The Endocrine System

Deborah Bates

Discover your body's health secret, the endocrine system, and 'twang' your way to sustainable health!

Paperback: 978-1-84694-503-8 ebook: 978-1-78099-697-4

Readers of ebooks can buy or view any of these bestsellers by clicking on the live link in the title. Most titles are published in paperback and as an ebook. Paperbacks are available in traditional bookshops. Both print and ebook formats are available online.

Find more titles and sign up to our readers' newsletter at http://www.johnhuntpublishing.com/mind-body-spirit

Follow us on Facebook at https://www.facebook.com/OBooks/ and Twitter at https://twitter.com/obooks